# MICROPROCESSORS IN PROCESS CONTROL

# MICROPROCESSORS IN PROCESS CONTROL

**J. BORER**

*Department of Mechanical Engineering,*
*Brunel University, Uxbridge,*
*Middlesex UB8 3PH, UK*

**ELSEVIER APPLIED SCIENCE**
LONDON and NEW YORK

ELSEVIER APPLIED SCIENCE PUBLISHERS LTD
Crown House, Linton Road, Barking, Essex IG11 8JU, England

*Sole Distributor in the USA and Canada*
ELSEVIER SCIENCE PUBLISHING CO., INC.
655 Avenue of the Americas, New York, NY 10010, USA

WITH 2 TABLES AND 263 ILLUSTRATIONS

© 1991 ELSEVIER SCIENCE PUBLISHERS LTD


**British Library Cataloguing in Publication Data**

Borer, John
    Microprocessors in process control.
    1. Process control. Applications of microprocessors
    I. Title
    670.4275

    ISBN 1-85166-578-1

**Library of Congress Cataloging in Publication Data**

Borer, J. R.
    Microprocessors in process control/by J. Borer.
    p. cm.
    Includes bibliographical references and index.
    ISBN 1-85166-578-1
    1. Process control.    2. Microprocessors — Industrial applications.
    I. Title.
    TS156.8.B664 1991
    629.8'95416—dc20                    90-22858
                                                     CIP

Phototypesetting by Tech-Set, Gateshead, Tyne & Wear.
Printed in Great Britain at the University Press, Cambridge.

# *Preface*

The microprocessor has revolutionised many fields of industrial and business activity over the last decade, and industrial measurement and control is no exception. Equipment systems now available are very much more reliable and have almost 100% availability. They also offer scope for implementing many control strategies which, though they have been theoretically possible for many years, it has not been practicable to implement until now. The reasons for this are firstly digital computing power which became available in the 1960s when DDC systems were introduced, and secondly distributed processing which is the product of microprocessor technology.

The skills required by instrumentation and control engineers in the process industries are more diverse than in many other disciplines. To this must now be added considerable knowledge of electronics and communication technology, if they are to be able to understand and take advantage of these systems. The process engineer is responsible for the structure of the *control* systems as opposed to the *equipment* systems; he must understand much more about the equipment than was necessary with analogue systems, if their advantages are to be fully exploited. This book is written in the hope that it will provide an introduction to the subject matter for these two groups of engineer at a level which is both acceptable to them and also adequate for them to comprehend the technologies involved. It is not written for those who aspire to design the equipment systems themselves, though it may serve as introductory reading even for this group. There are many excellent books on each of the technologies which contribute to these systems, written by engineers who have much deeper knowledge of a specific technology than the author of this book.

The first part of the book describes the technology of measuring, in an industrial context, the most common and therefore most important variables: pressure, level, flow and temperature. Part 2 of the book endeavours to provide a minimal basis in the established techniques by which process plant is regulated and controlled. These have not yet changed greatly as a result of the much greater equipment flexibility; in the immediate future they will certainly begin to do so. The pace of change of such technologies is, however, by its nature, much slower than that of the electronics technologies which have revolutionised the equipment systems. The third part of this book sets out to describe how these microprocessor-based equipment systems function and how they are constructed from the standard components available in 'chip' form. The later chapters of Part 3 introduce the reader to the problems of transfer of data within these systems and describe how security of operation is built into these systems to provide the reliability and availability necessary. Finally Part 4 reviews the way that measurement and control strategies can be (yet sometimes are not) implemented using these equipment systems.

JOHN BORER

# *Acknowledgements*

Thanks are due to the following for permission to reproduce previously published material:

American Meter Co. Inc. for Fig. 4.35;
British Rototherm Co. Ltd for Fig. 6.2;
Ferranti plc for Fig. 7.14;
Flo-tran Inc. for Fig. 4.31;
Foxboro Great Britain Ltd for Figs 2.5, 7.5, 7.7, 7.8 and 7.9;
Gervase Instruments Ltd for Fig. 4.19;
KDG Flowmeters for Fig. 5.13;
Kent Process Control Ltd for Figs 4.11 and 4.24;
Kistler Instruments Ltd for Fig. 2.9 (left);
Measurement Technology Ltd for Fig. 7.20;
Moore Products Co. (UK) Ltd for Fig. 4.30;
The Open University for Figs 2.6 (left), 2.7, 2.9 (right), 4.23, 4.25, 6.9, 6.10, 6.17, 6.24, 7.13 and 7.15;
Rosemount Engineering Co. Ltd for Figs 6.19, 6.30 and 6.31;
Taylor Instrument Companies for Fig. 4.5;
Thorn EMI Datatech Ltd for Fig. 2.6 (right);
Voest-Alpine AG for Figs 2.10-2.12, 3.5, 3.6, 3.19-3.23, 8.3-8.10;
Whessoe Systems and Controls Ltd for Fig. 3.8.

# Contents

**Part 2**

# PART 1

# CHAPTER 1

# *Principles of Industrial Measurement*

## 1.1 GENERAL

In order to operate chemical plant processes, e.g. chemical reactions, petroleum distillation, etc., it is essential to know the values of physical states of the process fluids, such as pressure, temperature and density, as well as rates of flow and often analytical data. Industrial instruments have been developed to measure all these parameters and in turn the instruments themselves depend on physical laws. Before we can use any tool (and instruments are tools for measuring) we need to know its capability. It is necessary to define limits of performance for any measuring instrument or system, and before we can do this the terminology used needs to be defined.

## 1.2 INSTRUMENT PERFORMANCE

It is important to determine with what precision measurements can be made using the instrument or system, but this will depend on many factors. Because of slack in linkages, friction and many other imperfections, repeated measurements made with the same system will only give the same result within a certain *error band*. This limitation on performance of a measuring system is referred to as *repeatability*. No matter how repeatable the results there will be a limit on the *resolution* with which they can be indicated or recorded. The measuring system will have a range or *span* over which it can work, and ideally a graph of the relationship of measured variable to instrument indication (or recording) will be a straight line (Fig. 1.1). In fact this will never be the case, and *accuracy* will be defined as the limit of confidence which can

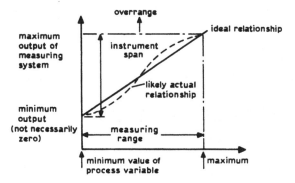

**Fig. 1.1.**

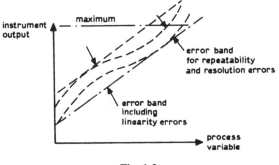

**Fig. 1.2.**

be placed in a measurement, taking all the factors into account
(Fig. 1.2).

## 1.3 RANGEABILITY

Any industrial measurement system should give information of
sufficient accuracy to facilitate control of the process operations over a
*range* of operating conditions. It is often forgotten, however, that many
of the causes of error are related to the maximum of the measuring
range. Manufacturers usually quote errors in terms of FSD (full scale
deflection). If a process variable is to be measured it is implied that it
varies in the course of normal process operation; to allow for such
variation the range of the measurement system will normally be selected
so that the normal operating value of the variable represents about 70%

of FSD. Thus, if typically, a range of process variable of 3 to 1 is to be measured and the system accuracy is $\pm 1\%$ FSD, then the errors to be expected at the lower end of the range will be

$$\left(1 \times \frac{1}{0.70} \times \frac{1}{3}\right) = \pm 5\%$$

## 1.4 ASSESSMENT OF ERRORS

It is easier to say how the performance of a measurement system is determined, than to determine it in practice. Whilst the instrument technician at a refinery or chemical plant will rarely, if ever, be asked to carry out such an evaluation experimentally, it is, nevertheless, essential that he should understand how this is done. Sometimes the errors caused in the different ways outlined above will cancel one another out; sometimes they will add up and so reinforce each other. Thus the actual error which occurs in any particular measurement is *randomly* determined; only the *probability* that the error will be greater or less than a certain size can be determined. Therefore the accuracy of a measurement system is always quoted in statistical terms, that is the size of error which has, say, a 90% probability of occurring; it cannot be quoted in any other way.

To establish this statistical data, experiments must be made repeatedly with the measuring system under test, so that the error is found on a sufficient number of occasions to allow the data to be reliably grouped; the probability of the occurrence of errors of different sizes can then be evaluated. This is obviously a very time-consuming method.

## 1.5 CALIBRATION

Industrial measuring devices and systems must be robust and easily maintainable, and to some extent accuracy is sacrificed to these ends. Any experiment to assess the error of measurement requires that there exists some other means of establishing the true value of the process variable being measured. Since such experiments are carried out under laboratory conditions, a more accurate instrument is often available so that the measurements can be made simultaneously on this and the industrial instrument under test. Such high accuracy instruments are known as *substandard*, and are usually at least one order of magnitude better in terms of accuracy (that is ten times more accurate). If such substandard instruments are not available, some fixed physical phenomenon, such as the boiling point of a liquid, must be used to evaluate the accuracy, at least at certain fixed points on the instrument range.

## 1.6 ZERO AND SPAN

As the range of the process variable will be different for each individual application, industrial instruments are almost always made in such a way that the *span* can be adjusted to suit each application within a wide range, so that sensitivity, accuracy and other sources of error are proportional to the span. In the course of normal operation the span of the instrument can change owing to vibration, heat, physical blows or any number of other causes, as can the zero setting. These two adjustments must therefore be checked (and if necessary corrected) frequently, often whilst the instrument remains in its installed position on the plant. Obviously it is not possible in most cases to use substandard instruments or other laboratory techniques, and such 'on plant' checks are usually made by 'injecting' known test inputs using special test equipment.

## 1.7 DRIFT

A very important aspect of the performance of industrial measuring systems is their propensity to 'drift'. Either span or zero may change gradually because of the ageing of components (this is very important in the case of electronic equipment) or other forms of slow deterioration. Such drifting is common in new equipment and for this reason span/ zero checking should be carried out more frequently immediately after installation. However, if an instrument should be found to drift continuously, long after its installation, it should be returned to the manufacturer as unsatisfactory. Unfortunately this type of failure often goes undetected and manufacturers rarely provide data on drift as part of performance specifications.

## 1.8 RESOLUTION

This term applies to the precision with which the measurement can be displayed, recorded, or logged. If the measuring system presents the results of the measuring process to the operator in the form of an indicated figure, then the size and length of the indicator scale will inevitably limit the size of the smallest unit which the human eye can 'discriminate'; however, there would be little point in supplying a larger indicator if the size of unit which could then be read is smaller than the limit of accuracy of the measuring system itself. Thus, the indicator or recorder should have a scale size which is consistent with the limits of accuracy of the measuring system.

A different situation arises in the case of digital displays which are becoming very popular; discrimination depends on the number of digits used to represent the measured value, regardless of the size of the display, and the precision of the electronic circuitry which converts the measurement into the digital quantity displayed (or recorded).

## 1.9 HYSTERESIS

A common cause of error in many measurement systems is hysteresis, caused by friction or by any one of a multitude of directional effects in mechanical, pneumatic, hydraulic or electrical mechanisms. These result in a different measurement being obtained if the process variable has increased to the measured value from some previous lower value, or alternatively has decreased from some previously higher value.

## 1.10 DIRECT/INDIRECT MEASUREMENT

For the most part the techniques used in industrial measurement are inferential or indirect. For instance, in order to measure the temperature of a process fluid, the pressure of a liquid or gas sealed into a metal container may be measured, and the temperature 'inferred' from the pressure according to certain known relationships. Some techniques, however, are 'direct', e.g. the measurement of flow rate of a fluid by the positive displacement technique, in which the meter actually transfers a 'package' of process fluid from one place to another, depending on mechanical seals to prevent any of it returning. Such techniques are usually more accurate though much more expensive than inferential techniques.

## 1.11 SAMPLING

It is necessary to ensure that the measurement made, though it may be perfectly accurate, is representative of the measurement required. For instance, the flow rate or temperature of a fluid may vary across the diameter of the pipe if the flow is not sufficiently turbulent to ensure good mixing. Again, even if flow is sufficiently turbulent, the flow rate or temperature may vary with the passage of time. Temperature or pressure may vary at different places in a large tank either in a regular manner, and therefore predictably, or in a random manner, and therefore unpredictably. In such cases it may be necessary to make more than one measurement in order to determine the mean value of the variable. If the 'distribution' of the variable in time or position is predictable (as, for instance, the distribution of fluid flow across a pipe) this may not be difficult: if, however, the distribution is random it will only be possible to make a number of measurements (as many as is practicable) and hope that their average is a close approximation of the true mean value of the variable. The probability that this is the case increases with the number of such measurements made and decreases with the differences in the measurements across the distribution in time or position. The calculation of a suitable tolerance is statistically based and will depend upon the degree of certainty which is considered adequate (there can never be absolute certainty in such cases); for instance, there may be a 90% certainty that the average of the measurements is within $\pm 1\%$ of the true mean value.

# CHAPTER 2

# *Measurement of Process Pressure*

## 2.1 INDIRECT PRESSURE MEASUREMENT

The commonest instrument used to measure pressure in a process plant is the pressure gauge. This is normally mounted close to the point of measurement and connected to it by an 'impulse' pipe, so that the process fluid is brought up to, and in most cases, into the gauge. Elastic deformation of a suitable measuring element is translated into motion of a pointer across a scale by mechanical gearing.

The commonest form of measuring element used is the 'Bourdon' tube which is a tube formed into an arc of a circle as shown in Fig. 2.1. Increasing internal pressure causes the tube to deform elastically in such a way that it straightens out. Different cross-sectional shapes are used for different pressure ranges and applications.

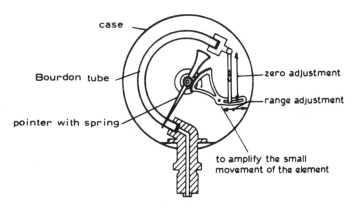

**Fig. 2.1.**

7

The Bourdon tube exhibits hysteresis error and the zero is often indeterminate and may change with use and ageing. Wear in the linkage also produces deadband error and expensive gauges use jewelled bearings; however, the most serious source of error is temperature variation. The Bourdon tube changes shape in response to variations in temperature unless made of a material which has a near-zero coefficient of expansion, such as Ni-Span-C. Cheaper guages are made from copper or brass, whilst for very high pressures, hardened-steel solid drawn Bourdon elements are used.

The form of Bourdon tube shown in Fig. 2.1 is known as the 'C' type, but other forms are used to obtain greater movement from the same pressure change (Fig. 2.2). The movement of a Bourdon element is proportional to the angle through which it is bent: the 'C' form is bent through approx. 200° whilst a spiral or helical element can be bent through 1000°. The helical form is used for higher pressures as the radius of curvature, and therefore the stress, is uniform over the whole length of the element, which is not the case for the spiral form.

The Bourdon element is very sensitive to vibration and is prone to fatigue failure if subjected to fluctuating pressure. It is also prone to corrosion in some cases owing to the fact that the process fluid enters the element. All these problems can be overcome by using a diaphragm element. The diaphragm can be protected against overpressure by limiting its possible movement as shown in Fig. 2.3; it can also be effectively protected against corrosion by coating with silver or plastic.

The capsule measuring element (Fig. 2.4) consists of two metal diaphragms soldered together to form a closed capsule. When the pressure rises the capsule expands and thus operates the pointer mechanism. Capsules are used to measure absolute pressures because

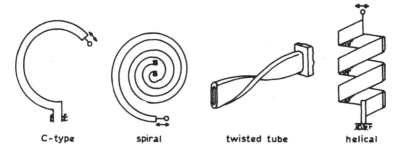

C-type        spiral        twisted tube        helical

**Fig. 2.2.**  Bourdon tubes.

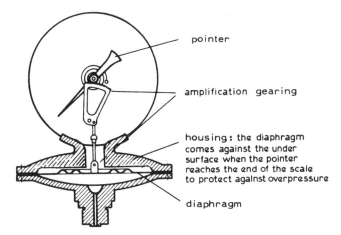

pointer

amplification gearing

housing: the diaphragm comes against the under surface when the pointer reaches the end of the scale to protect against overpressure

diaphragm

**Fig. 2.3.**

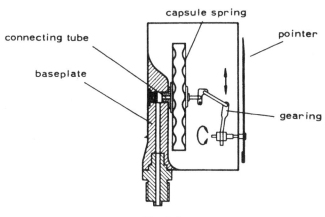

capsule spring

connecting tube

pointer

baseplate

gearing

**Fig. 2.4.**

they are easily 'evacuated' and can be made very sensitive to small pressure changes.

Stiff diaphragms are 'stacked' to produce measuring units with much greater movement than a single capsule. One of the best known measuring devices which uses stacked diaphragms is the 'Barton' meter (Fig. 2.5). This unit is used to measure small differential pressures where the background pressure is high. It can withstand overpressures because each diaphragm is shaped to fit into the next in such a way that

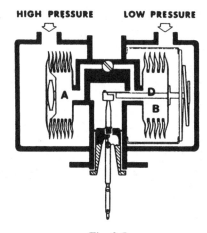

**Fig. 2.5.**

when the 'stack' is fully compressed it is 'solid' and resists any further deformation.

The two stacks are filled with liquid, and increasing pressure in the high pressure chamber compresses stack A, thus displacing liquid into and expanding stack B and moving the pointer arm D. Reduction of pressure in the low pressure chamber has the same effect, whilst an increase in this chamber or a decrease in the high pressure chamber moves the pointer in the opposite direction. The spring C adds to the elastic resistance of the diaphragm stacks to the applied pressure and thus allows the instrument to be calibrated to respond over the desired range.

Corrugated bellows are often used in place of stacked diaphragms as they can be made more easily by a single pressure-forming operation (stacked diaphragms involve soldering each pair of diaphragms).

## 2.2 MEASUREMENT BY ELECTROMECHANICAL METHODS

All the measuring elements described so far depend on the elastic properties of metals to provide a force which is proportional to the process pressure: in every case the elastic *movement* of the metal element is measured, in many cases after amplification by mechanical gear and lever mechanisms. An electrical response can be used very easily to cause a pointer or recorder pen to move across a scale or chart. Such instruments produce an electrical current or voltage in response to pressure, by virtue of a change of capacitance, inductance or resistance.

All electrically conductive materials show a change in resistance if stressed, and this 'piezo-resistive' phenomenon is used to produce the 'strain gauge'. Strain gauges are fixed to a diaphragm or other mechanically deformed measuring element. Instead of measuring the movement under pressure the change in resistance of the strain gauge is measured; 'strain' being the change of length of an elastic element which accompanies the stress set up by pressure on the diaphragm or other mechanical measuring element. The movement which is necessary to cause an adequate (for measuring purposes) response from a strain gauge is very much less than that which could be properly measured by mechanical means. This enables 'stiffer' and therefore more robust measuring elements to be produced, which is particularly important when considerable 'over-pressure' may have to be tolerated by the measuring element, or when measuring very large pressures or forces. Pressure being force per unit area, the measurement of force and pressure are almost the same.

Capacitance or inductance can be changed by small deflections of a sensing diaphragm, thus providing alternative ways of measuring the deflection by electrical means (see Fig. 2.6).

Using any of these methods it is necessary to translate the change in electrical resistance, capacitance or inductance into an electrical force capable of moving the pointer or pen of the indicating or recording device in the measurement system. This is achieved by using a 'bridge' as shown in Figs 2.7 and 2.8.

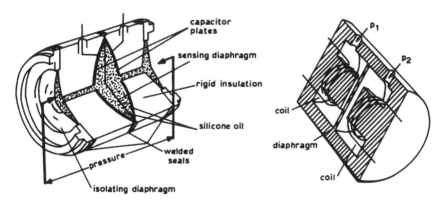

**Fig. 2.6.**

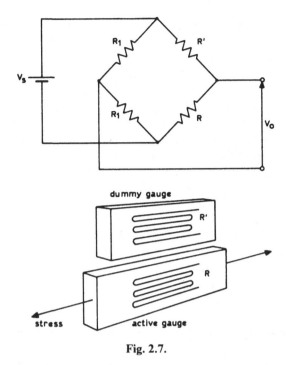

**Fig. 2.7.**

The 'bridge' operates on the principle that the input voltage is divided across the two fixed reference resistors in the left-hand 'leg' and across the two measuring resistors, capacitors or inductances in the right-hand leg, in proportion to the values of these components. Thus, the two reference devices are of equal value, as are the two measuring devices at zero reading; as the pressure changes, however, the two measuring devices are no longer of the same value and the voltage between them changes and is no longer the same as that between the two reference devices. This difference produces a voltage, $V_0$, at the output which with power derived from the 'bridge' source drives a pen or pointer. This power source must be alternating in the case of inductance or capacitance bridges and can be either direct or alternating in the case of resistance (strain gauge) devices.

## 2.3 MEASUREMENT BY ELECTRICAL METHODS

If a quartz crystal is 'squeezed' between two parallel faces, it responds by generating a voltage across those faces in proportion to the pressure

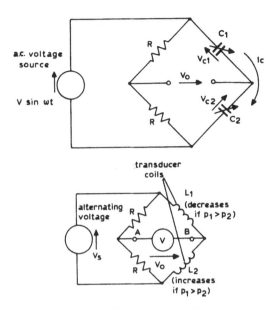

**Fig. 2.8.** (a) An a.c. bridge for differential capacitance measurement. (b) An a.c. bridge circuit used to detect the relative changes in the inductances $L_1$ and $L_2$.

applied (Fig. 2.9). This 'piezo-electric' effect can be used to generate power to drive a pen or pointer directly by means of a special electronic circuit called a 'charge amplifier' without any mechanical motion at all. Thus, the piezo-electric sensing element is a true electrical device.

The electrical charge produced by a piezo-electric measuring element is proportional to the *change* in pressure applied, and, however perfect

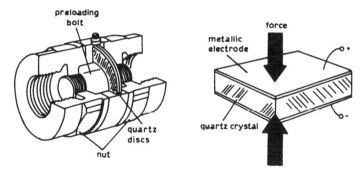

**Fig. 2.9.**

the charge amplifier, the charge will decay in time. For this reason it cannot be regarded as a viable method of measuring relatively steady pressures since no reliable zero can be established. It is used in fact to measure very rapidly fluctuating pressure where the fluctuation rather than the mean pressure is important (the mean can be established by a separate electro-mechanical instrument). Thus, piezo-electric sensing elements find little application in process operations.

## 2.4 SELECTION OF MEASURING ELEMENT

The most common form of pressure-measuring element used in plant processes is the Bourdon tube. Pressure gauges, i.e. indicating instruments mounted directly on the process pipe or vessel, almost invariably incorporate a Bourdon tube. The exceptions to this rule are found where the process fluid is highly corrosive, requiring the use of construction materials for the measuring element which are more suitable for a diaphragm than a Bourdon tube; or when vibration might be a problem, since a diaphragm is less liable to fatigue failure than a Bourdon tube in these circumstances. The diaphragm or 'Shaffer' gauge is also more suited to measuring fluctuating pressure for the same reason and can be made more sensitive than a Bourdon tube for measuring pressures lower than 1 bar. For highly corrosive applications the diaphragm can be coated with such materials as PTFE or silver. The Bourdon tube, however, in its simple 'C' form, and manufactured in a variety of materials from phosphor–bronze to chrome–molybdenum steel, is used to measure pressures from 1 bar up to at least 5000 bar.

In the spiral or helical forms the Bourdon tube is used in instruments which require greater movement than can be obtained from a 'C' form, e.g. circular chart recorders. This eliminates the need for a quadrant and pinion gear which is the cause of deadband error and lack of sensitivity in the cheaper 'C' element.

For measuring small pressures the sensitivity and lack of temperature dependence of the capsule element make it ideal, either in the single form, or, where greater movement is required, in the stacked form used in the Barton meter. The measurement of small differential pressures at high static pressures can be achieved in this way, and often bellows are used instead of stacked diaphragms.

Piezo-electric measuring elements are not normally used in process plant because they only measure the change in pressure and not the total pressure. Electro-mechanical elements do have the advantage that they can be made much stronger than elements which must distort enough to operate a linkage, and also that deadband error, which increases as pivots and bearings wear, is avoided. However, the precautions which must be taken to avoid risk from incendiary sparks is a serious drawback. This is less serious when electrical transmission of signals is used.

## 2.5 CALIBRATION OF SENSOR ELEMENTS

In general the 'direct' methods are used to calibrate sensors which operate on 'indirect' (inferential) principles. This is because the direct methods, i.e. 'U' tube or single leg manometers, deadweight testers, etc., are capable of higher accuracy whilst at the same time being too cumbersome for use on the plant. Measuring elements which depend on the stress set up in an elastic material (which is the case with all the elements described) must be checked periodically to see if their span or zero setting has 'shifted' owing to slight non-elastic behaviour of the material; this is the main disadvantage of indirect methods. Calibration checks will also be necessary if the instrument has been subjected to overload, sustained fluctuations or severe vibration in service. Calibration checks, as well as the original calibration, must be made by comparison of readings taken from the instrument under test against readings taken on a sub-standard instrument when both are subjected to the same pressure at the same time. The substandard must be 10 times more accurate, and this is achieved by using a device which is capable of greater accuracy and which can be expected to maintain that greater accuracy over long periods (if this were not the case, the sub-standard instruments would require calibration checking as often as the 'field' instruments). Periodically, and also in the event of any possible damage, the sub-standard instrument will have to be checked in its turn; this is usually achieved by sending it to an organisation that specialises in calibration because the precision required to achieve the even greater accuracy in the 'reference' standard device is normally beyond the capability of a plant instrument workshop.

The extent of calibration checks vary with the instrument and its application; a simple low-accuracy pressure gauge will have only zero calibration adjustment — it is assumed that the range (span) will not change outside the performance limits set (some cheap gauges may not even have zero setting adjustment). Nevertheless, calibration checks are usually necessary in order to ensure that the gauge is indeed still within calibration. Most instruments will have adjustable spans as well as zero settings; periodical checking will discover if the instrument is drifting in respect of either. Hysteresis is normally a feature of the instrument and errors from this cause would not normally change during use. However, deadband is often caused by slack in mechanical movements and can be expected to increase with use in instruments which rely on gears, pivots, etc. Linearity is usually a feature of the instrument, but may be changed by 'softening' of springs or bellows after considerable use.

Calibration checks on field instruments are normally a matter of comparing readings on the instrument under test and the 'substandard' at about 10 points up *and down* the range. For especially accurate 'field' measurement systems (which are fortunately rare) it may be necessary to adopt a 'statistical' approach and make a large number of such tests. From such data the 'norm' and the 'standard deviation' of repeated readings can be calculated, and these may be a better guide to the performance (and its possible deterioration) than a single set of results.

Deadweight testers are the most accurate calibration instruments in use for pressures above those which can reasonably be measured using a manometer. Pressure is provided by weights acting on a piston which fits very closely into a cylinder containing oil (Fig. 2.10(a)); the pressure developed in the oil is equal to the weight divided by the cross-sectional area of the piston, and this pressure is applied to the inferential instrument under calibration:

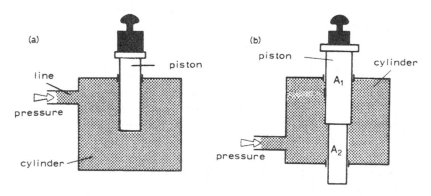

**Fig. 2.10.**

$$p = \frac{W}{A}$$

Friction between the piston and the cylinder is the only source of error and in a well-made instrument this is negligible.

To obtain very high pressures the piston is stepped as shown in Fig. 2.10(b) and a second gland added. This has the effect that the area over which the weight is distributed is the difference between the cross-sectional areas of the two sections of the piston, and thus the pressure generated by even quite a small weight can be very high indeed:

$$p = \frac{W}{(A_1 - A_2)}$$

The ring balance manometer comprises a tube bent into a circle and supported on a knife-edge pivot so that it can rotate (Fig. 2.11). Pressures $p_1$ and $p_2$ are isolated from each other by a partition on the one hand and liquid fill on the other. Because of the difference in pressure across the partition (which is fixed to the tube), a turning moment is applied to the tube; this is balanced by a turning moment produced by a counterweight and the angle of rotation is a measure of the pressure applied. For very low pressures the 'bell' ring balance gauge (Fig. 2.12) provides a solution to the errors caused by the stiffness of the tubes connecting $p_1$ and $p_2$ to the process.

## 2.6 SEALS AND PURGES

In many applications the process fluid cannot be allowed into the measuring element. There are many reasons for this; the fluid may solidify at the temperature of the measuring element, deposition may occur in the element or the fluid may be corrosive. In all these cases it may be necessary to interpose a 'seal' between the process and the measuring element. A typical seal is shown in Fig. 2.13.

The space between the seal and the measuring element is filled with a liquid that will not vaporise and all air and gas is removed by bleeding. Care must be taken that the seal and impulse lines are not subjected to changes in temperature to the extent that expansion of the fluid 'fill' (or contraction) would bring the bellows up to their limit stops; provided this precaution is taken and the 'spring rate' (or resistance) of the

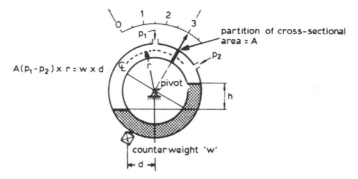

**Fig. 2.11.** Ring balance gauge.

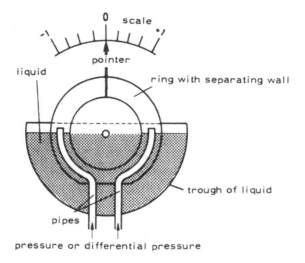

**Fig. 2.12.** 'Bell' ring balance gauge.

bellows is negligible compared with the pressure to be measured, the pressure will be transferred from the process to the 'fill' fluid. Such a seal would serve for a clean fluid that would be likely to solidify in impulse lines, provided the seal itself was well lagged. It would not be suitable for use with a dirty fluid which would allow sludge to gather in the bellows corrugations, or a process fluid which would deposit solid matter. For such applications a plain or corrugated diaphragm would replace the bellows. More likely, however, a purge system would be engineered for such an application.

The principle of a purge system is that a fluid (liquid or gas) is allowed

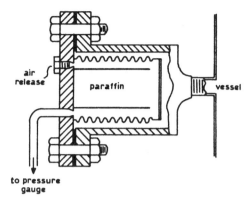

**Fig. 2.13.**

to flow into the process fluid through a narrow bore tube at such a rate that the process fluid itself is prevented from entering; thus if the process fluid is dirty, corrosive or likely to deposit solid matter on the measuring element, a purge will enable measurement to be made nevertheless. However, it is not always acceptable to introduce another fluid into the process line, so the purge technique cannot always be used. A purge system is illustrated in Fig. 2.14; the self-contained flow regulator ensures that the correct flow rate is maintained whatever the pressure in the process stream, provided the supply pressure in the purge line is at an appreciably higher pressure than the highest process pressure possible. The purge fluid may be a gas, such as air or nitrogen, or a liquid, such as water or paraffin.

## 2.7 DAMPING

One of the main dangers in measuring pressure is that surges may overpressure the instrument, or that rapid fluctuations may cause

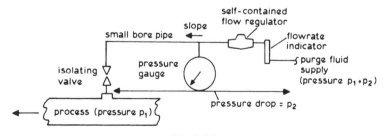

**Fig. 2.14.**

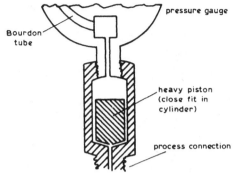

**Fig. 2.15.**

fatigue fracture in the measuring element. Such dangers are avoided by using dampers (or 'snubbers') such as the one shown diagrammatically in Fig. 2.15. The inertia of a relatively heavy piston fitted in a chamber between the process and the instrument in the vertical plane absorbs the energy of surges or fluctuations.

## 2.8 HYDROSTATIC HEAD CORRECTION

The measuring element will not normally be located at the same level as the pressure tapping in the process pipe or vessel, as considerations such as access for service and reading tend to dictate the location. If, therefore, the process fluid is a liquid, the hydrostatic head must be allowed for in calibrating the instrument; often this will mean that routine zero/span calibration checks will be made in the field, the substandard instrument being taken to the field instrument instead of the latter to the workshop. If the process pressure is relatively low, care is necessary in selecting a position for installation to ensure that the hydrostatic head correction is within the scope of the zero adjustment available.

## 2.9 PROCESS PRESSURE CONNECTIONS

Whilst the impulse piping from the pressure connection on the process pipe or vessel to the measuring instrument must obviously be rated to withstand the maximum pressure to which it may be subjected in the course of operation, there are many reasons why it may not be necessary or even desirable to follow the process piping or vessel specifications beyond the first isolating valve. For one thing, should a leak occur, the impulse line and instrument can be isolated and repaired without affecting the process operation, whereas this may not be possible in the case of the process pipework or vessels. Consequently the first isolating valve has a special significance; the outlet is taken as the boundary beyond which 'instrument' rather than 'piping' specifications apply. The valve should always be located as close to the tapping point as possible.

## 2.10 SAFETY

Measuring instruments vary in their capacity to withstand pressures above those which they are designed to measure; this will always be a feature of the design of the instrument. Instruments may on the other hand have to withstand 'overpressure' for a variety of reasons, the most obvious being that pumps have a higher 'no flow' head than that developed at 'design' flow rate. It is not always possible to buy an instrument that will withstand the maximum pressure that can, under fault conditions, occur in the process; this will usually be set by a safety valve in the process line. It is essential, however, that the instrument should be able to withstand the highest pressure that can occur in *normal* operation.

Even if the instrument is never subject to overpressure, the measuring element (which is normally the weakest part) may still fail because of a manufacturing fault or, after considerable use, metal fatigue. For this reason the case in which the measuring element is housed will be provided with a 'blow-out disc' or some other means to relieve the flow of process fluid which could result from a catastrophic failure of the element. Care should be taken when installing the instrument not to obstruct such relief devices and also to see that they are directed away from positions where people are likely to stand, particularly if the process fluid is toxic, corrosive, hot, etc. The glass in recording or indicating instruments will be blown out if these relieving devices are restricted, causing another hazard. Finally, there must always be an isolating valve between the 'process' and 'instrument' piping because of these factors — since the measuring elements must operate within the 'elastic' region they cannot be made not to rupture under any circumstances. For such absolute safety, strain-gauge methods must be used.

# CHAPTER 3

# *Fluid Level Measurement*

## 3.1 DIRECT METHODS OF MEASUREMENT

The simplest method of measuring the liquid level in a vessel is the sight glass (Fig. 3.1). However, a sight glass behaves exactly the same way as a manometer and so the level in the glass is only an accurate measure of the level in the vessel provided the density of the liquid in the glass is the same as that of the liquid in the vessel. Therefore, when the temperature of the liquid in the vessel is much higher than ambient, as in the case of a boiler drum, or when it is much lower, as in the case of a cryogenic plant, the level will not be correctly measured and a correction will have to be applied.

The other great drawback to the sight glass even for local indication is the fragility of glass. On the other hand, because it is such a simple device, and therefore there is little that can go wrong with it, sight glasses are required by law on such plants as steam-raising boiler drums. To overcome the attendant risk of breakage, and consequent leakage of

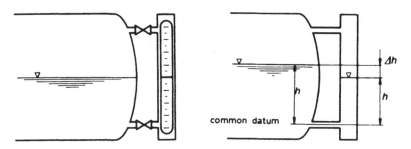

**Fig. 3.1.**

22

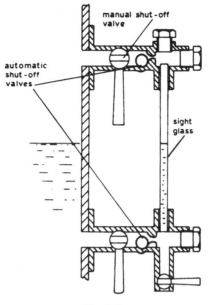

**Fig. 3.2.**

high-temperature liquid at high pressure, automatic shut-off valves are built into the sight gauge as shown in Fig. 3.2.

To avoid the use of glass and to enable the measuring tube to be thermally insulated, magnetic followers are sometimes used with tubes made of non-magnetic material such as stainless steel (see Fig. 3.3). Strictly speaking this is not a sight 'glass' but the principle is the same.

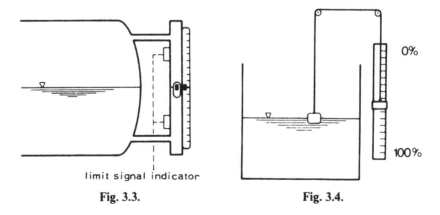

**Fig. 3.3.** **Fig. 3.4.**

The float and pulley method of level measurement has been used in various forms for a very long time (Fig. 3.4). The method in its simplest form depends on the weight of the float being counterbalanced by a weight at the other end of the pulley system (at the indicator), thus putting the wire (rope or tape) into tension. At the same time the full weight of the float is partly diminished by its buoyancy in the process liquid. Thus, to reduce errors which would arise owing to changes in the density of the fluid (and consequent changes in buoyancy) and also to friction in the pulleys, it is necessary for the float to have as large a cross-sectional area as possible, so that a small rise or fall in its position in the liquid will represent a large change in buoyancy force. Another source of error arises from the transfer of weight of the rope (wire or tape) from the float side of the system to the indicator side. A more accurate system is shown in Fig. 3.5; the rope is wound onto a drum as the float rises, and tension in a spring inside the drum reduces as the drum winds in the rope. The tension in the spring can be made to equal the weight of the rope at all positions, either by the natural change in tension as the spring winds up or unwinds, or by changing the diameter of the drum and thus the length of the moment arm, as the rope coils onto the drum.

Very accurate measurement of the liquid level is required for inventory purposes in large storage tanks — accuracy of better than ±2 mm *irrespective of the range of measurement*. This is not attainable in a mechanism which depends on the buoyancy of the float to provide the force to overcome pulley friction, etc.; a servo mechanism is used to supplement, if not replace, this driving force. A simple form of servo-

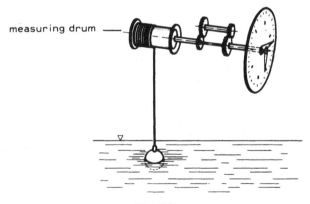

measuring drum

Fig. 3.5.

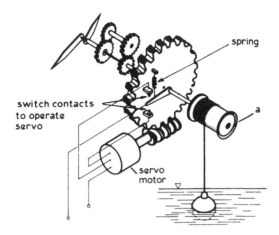

**Fig. 3.6.**

assisted gauge is shown in Fig. 3.6. Part of the weight of the float and rope is supported by buoyancy and the rest by the spring. When the level changes this equilibrium is disturbed, with the result that one or other of the switch contacts is 'made', causing the servo motor to drive the float either up or down until equilibrium is again restored and the motor circuit broken. Because the motor, and not the float, provides the motive force, a large indicator can be used and also a remote indication system if desired.

The problems of rope/wire weight and the force required to drive an indicator mechanism can be overcome in another way, which provides even more accurate indication. A metal plate is lowered on the end of a metal tape until it is almost touching the liquid surface; the electrical capacitance between the liquid and this plate is measured, and a servo motor drives the plate up or down to maintain capacitance, and hence the distance from the liquid surface, at a constant value to within ±1 mm. The indicator unit is mounted on top of, or at the side of, the tank, and this type of measuring system is used for accurate liquid level measurement in very large storage tanks. The servo unit is mounted in the indicator housing and the stainless steel tape runs over pulleys (as shown in Fig. 3.8) inside sealed pipes, thus making it suitable for use with tanks which are under slight pressure, e.g. petroleum storage tanks.

Servo gauge systems are electrically operated and precautions must be taken to ensure that they are safe to use in locations where there is a risk of explosion or fire from petroleum vapour. To achieve this, the servo mechanisms are invariably enclosed in flame-proof or explosion-proof enclosures; therefore, it is not possible to carry out maintenance or adjustments to the servo mechanisms in position, for which reason a special calibration 'rig' is essential even if the measurement range is 50 m or more. However, a limit stop is normally provided which gives a datum against which to compare the reading of the gauge; since it is the distance from this datum of the liquid surface which is being measured, there is normally no reason for the system to be out of calibration (it either works or it does not). Accuracy is effected by certain adjustments in the servo mechanism, and it is these adjustments which cannot safely be made in position, and which necessitate the gauge being removed to a workshop and installed in a special calibration rig.

Figure 3.7 shows a tanktop-mounted gauge which uses a magnetic follower so that the internal enclosure of the servo system can be isolated from the process environment. Figure 3.8 shows some details of a tankside gauge system using pulley and stainless steel tape.

## 3.2 INDIRECT MEASUREMENT BY HYDROSTATIC HEAD METHOD

Since pressure is force per unit of area, the pressure at the bottom of a tank of liquid depends on the height of the liquid above. The bottom of the tank is called the datum of measurement, and the level can be

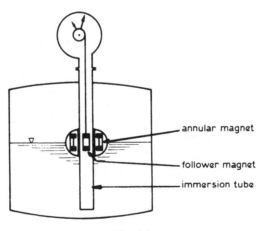

annular magnet

follower magnet

immersion tube

**Fig. 3.7.**

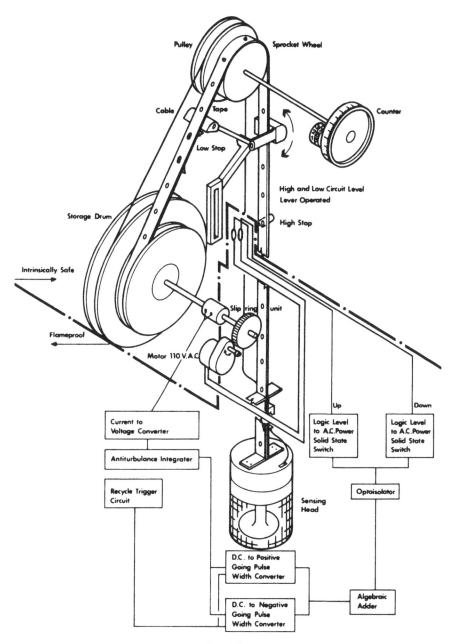

**Fig. 3.8.**

measured as the height of the liquid above this datum by inference from the pressure *at the datum*. The level in an open vessel is proportional to the gauge pressure at datum (Fig. 3.9). The pressure in a closed vessel is proportional to the differential pressure between the datum and the space above the liquid (Fig. 3.10). Note that the level in the vessel is not normally the same as that of the fluid in the manometer; the manometer fluid may be of different density and temperature. The manometer measures hydrostatic pressure, *not* level.

The datum of measurement does not have to be the bottom of the tank and in fact rarely is. However, measurement of the level is always relative to the datum level, which can be anywhere at or above the bottom of the tank. The level cannot be measured below the datum, which is fixed by the position at which the measuring gauge is connected into the vessel (Fig. 3.11).

Liquids with low boiling points are those with boiling points which lie below the ambient temperature (e.g. liquid nitrogen). In order to measure such liquids the tank is insulated and its temperature maintained below the ambient temperature. The lower impulse line is not insulated, and provided the vapour pressure at ambient temperature is higher than the hydrostatic pressure at the datum level, it will fill with vapour not liquid (Fig. 3.12). As any liquid entering would affect the measurement, the impulse lines are at a slight slope to the tank. It may

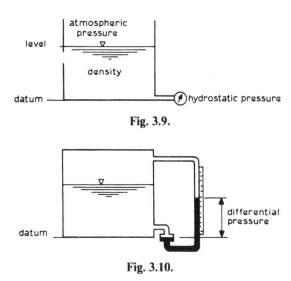

**Fig. 3.9.**

**Fig. 3.10.**

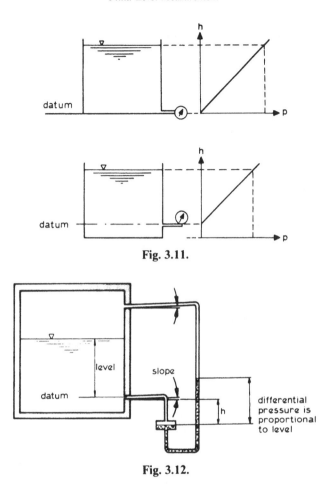

Fig. 3.11.

Fig. 3.12.

be necessary to heat the lower impulse line to make sure it contains no liquid.

Condensation chambers are essential when measuring the liquid level of boiling liquids, because the vapour will condense in the upper impulse line which will be cooler than the inside of the tank (Fig. 3.13). The cross-sectional area of the condensation vessel is made large so that the level of the condensed fluid will not change significantly when the manometer liquid rises with the level in the tank. Any such rise will result in condensate draining back into the tank, whilst any fall will quickly be made good by further condensation. In fact, vapour will be

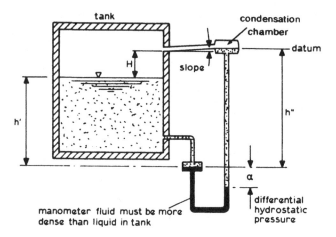

Fig. 3.13.

condensing all the time and draining back into the tank. The measurement datum is the surface of the condensate in the condensation vessel, as this does not vary. The level of the liquid surface in the tank *below* the datum is (Fig. 3.13)

$$H = h'' - h'$$

and is proportional to the differential hydrostatic pressure $\alpha$. Measurement of $\alpha$ must take into account the relative densities of the manometer fluid and the condensed process fluid.

If the process fluid is a viscous liquid it will not flow easily into impulse lines, and if it tends to crystallise or solidify it is likely to block them. In such cases, and for slurries, seals will be used at each pressure tapping point as described in Chapter 2 (see also Fig. 3.14).

## 3.3 INDIRECT (INFERENTIAL) MEASUREMENT BY BUOYANCY TECHNIQUES

The *apparent* weight of a body which is immersed in a fluid is less than the actual weight by the weight of the fluid displaced. In the sensor shown in Fig. 3.15 the 'apparent' weight of the float is supported by the spring, which extends or contracts as the level of fluid in the vessel, and hence the apparent weight of the float, changes. The range of

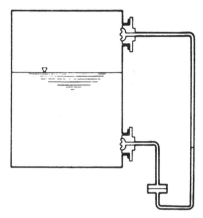

**Fig. 3.14.**

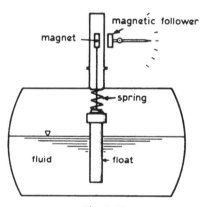

**Fig. 3.15.**

measurement is determined by the spring rate, and motion is amplified. The magnetic follower removes the need for a sealing gland. This is an inferential method of measurement, the apparent loss of weight of the float as the level rises depending on the density, as well as the volume, of the fluid displaced. Therefore, unlike the float/wire/pulley method the meter must be calibrated for a particular fluid, and changes of temperature will cause errors of level measurement. Other versions of this sensor use a torque tube and lever arm in place of the spring to transmit force (not motion) to a signal transmitting unit (see later).

## 3.4 THE USE OF 'BUBBLERS'

An alternative to the use of seals is the 'bubbler'. A gas is allowed to flow slowly through a vertical tube into the tank as shown in Fig. 3.16. In order for this to happen the gas must be at the same pressure as the hydrostatic pressure in the tank at the point at which it comes out of the bubbler tube; the bottom end of the tube therefore determines the datum of measurement. A self-contained flow-rate regulator is used to control the rate of gas flow so that it is the same at all levels of the liquid in the tank. The bubbler has the advantage that the measuring element can be located at any level; its position no longer determines the datum, as it is gas pressure not liquid hydrostatic pressure which is measured.

The gas can be instrument air or an inert gas such as nitrogen where necessary; it must not react with the liquid whose level is being measured. Instrument air must never be used to measure the level of petroleum in closed tanks because of the danger of forming an explosive mixture. As nitrogen is often used to 'blanket' such tanks, a bubbler system may be very appropriate.

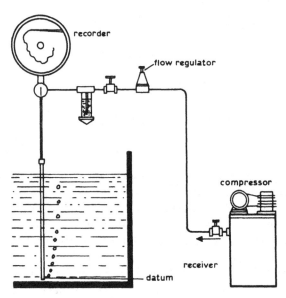

**Fig. 3.16.** Air purge installation with compressor.

## 3.5 THE USE OF 'LIMP' DIAPHRAGMS AND CAPSULES

The hydrostatic pressure at the datum level in a tank of liquid can be measured by a limp diaphragm sensing unit as shown in Fig. 3.17. The large 'pressure space' in the sensing element communicates via a small diameter capillary impulse tube with a bellows pressure sensing element of small volume, which may be mounted in any position, limited only by the length of the capillary. The gas (usually air) trapped in the pressure space is compressed until its pressure equals that of the hydrostatic head, the change of volume being accommodated by the 'limp' rubber diaphragm. The bellows element measures hydrostatic pressure, and the gauge has to be calibrated to take account of the density of the fluid, and is subject to errors due to temperature effects (change of density).

The capsule type gauge shown in Fig. 3.18 operates in the same way

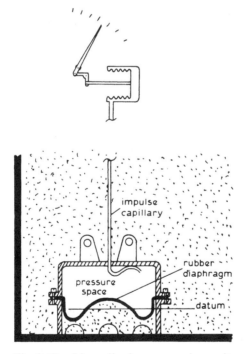

**Fig. 3.17.** Limp diaphragm sensing unit.

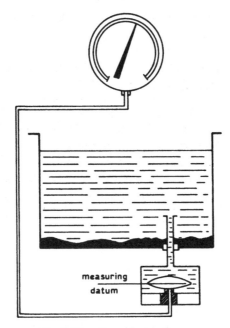

**Fig. 3.18.**  Capsule type gauge.

except that the hydrostatic pressure is opposed partly by the stiffness of the sensing capsule itself. Calibration must take this into account, and whereas the 'limp' diaphragm of the previous gauge could be replaced without affecting calibration, the capsule could not.

## 3.6 MEASURING LEVELS ELECTRICALLY

The measuring techniques described in this section depend on the different electrical properties of the fluid whose level is to be measured. They can only be used when these properties are suitable.

### By Capacitance
The insulated sensor (capacitive electrode) forms a capacitor with the container (Fig. 3.19). The dielectric constant depends on the liquid being measured. If the level of the liquid changes then the capacitance also changes proportionately, provided that the dielectric constant of the liquid is much greater than that of air.

A high-frequency voltage of constant frequency is used to measure

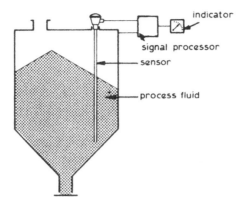

**Fig. 3.19.**

the capacitance. The current flowing through the capacitor is proportional to the level. This high-frequency current is converted to a standard output signal in the signal processing equipment.

The prerequisite for this measuring process is above all a stable dielectric constant. This is not likely to be the case if the liquid is a mixture or two-phase, and in particular the presence of small quantities of water in petroleum causes serious errors in this type of instrument. When measuring water-based liquids, very small quantities of dissolved solids cause serious errors. This type of measurement is most suitable for measuring the depth of powders in storage bins, as the dielectric constant is usually stable, and few other methods are available.

*By Conductivity*
Two electrodes, immersed in the vessel, are at constant voltage (Fig. 3.20). The area of cross-section between the electrodes increases as the level rises. The electrical resistance is reduced in the same ratio as the cross-section increases. The result is an increase in current, which is measured. The conductivity of most liquids changes with temperature, and the instrument must be calibrated for the particular liquid and compensated for temperature variation.

## 3.7 MEASURING LEVELS WITH SOUND PRESSURE

This technique is based on the measurement of the transit time of sound waves transmitted to, and reflected from, a surface. The velocity of sound is constant at constant temperature for any fluid. The time elapse

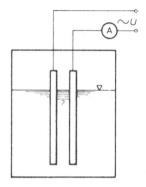

**Fig. 3.20.**

between transmission and reception is measured and used to determine the level.

The equipment should be mounted on the vessel so that the sound waves hit the material at right angles to the surface of the liquid, to ensure that they are reflected back to the receiver and not lost (Fig. 3.21). The method has the advantage that no contact is necessary between the sensor and the process fluid, and also that measurement is not dependent in any way upon the properties of the process material (density, dielectric constant, conductivity, etc.) provided a reflection can be obtained. This depends on the density of the liquid or solid process material being much greater than air. The technique is dependent upon the speed of sound pressure through air, and compensation must be made for the temperature of the air, since the speed of sound pressure varies slightly with temperature.

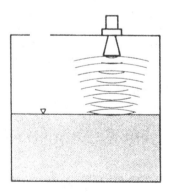

**Fig. 3.21.**

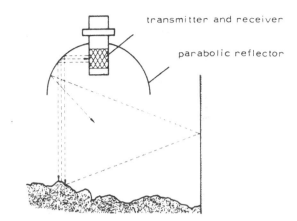

transmitter and receiver

parabolic reflector

**Fig. 3.22.**

If the surface of the liquid is very disturbed, or if it is a powder surface which is being measured, some of the reflected sound energy will be 'scattered' as shown in Fig. 3.22. This will tend to confuse the receiver since scattered sound pressure waves travel further than directly reflected energy and thus take longer in transit. To overcome this difficulty both transmitter and receiver can be mounted in a parabolic reflector which will reject reflected sound waves to a large extent.

The frequency of sound used depends upon the distance of the transmitter/receiver from the surface of the liquid or powder and the range of measurement; the shorter these distances the higher the frequency used. For large measuring ranges the frequency used is often low enough to be within the range of response of the human ear and can easily be interfered with by other extraneous noises.

Apart from faulty temperature compensation, the calibration of the instrument cannot alter since it is not dependent in any way upon the properties of the fluid or powder, the level of which is being measured. This means that a factory-calibrated instrument can be used without 'proving' equipment. Dirt or condensation, scatter and interference will affect accuracy, but not calibration.

## 3.8 MEASURING LEVELS WITH RADIATION

This technique depends on the fact that gamma-ray emission from a nuclear source is reduced in intensity as it passes through a solid or

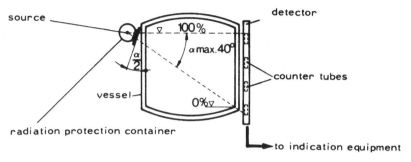

**Fig. 3.23.**

liquid in proportion to the density and the path length. A narrow beam of gamma radiation penetrates through both walls of the process vessel and the contents; the residual strength received at the detectors depends upon the path length through the contents which varies with the level in the vessel (the path length through the walls is constant and can be 'calibrated out'). Thus, provided that the liquid is much more dense than the air or gas in the space above and that the vessel walls are not excessively thick, the level can be determined from the variation in the rate of counting of gamma particles of the counter tubes.

The radiation protection container (with lead jacket) in which the radioactive source is situated, is mounted on the side of the vessel (Fig. 3.23). The detector is fixed on the opposite side of the vessel. A number of Geiger Müller Counter Tubes are installed, according to the level to be measured. The lowest counter tube will record the lowest count rate as the gamma particles reaching it will have passed through the greatest length of process liquid. The count rate will increase progressively towards the top until a level is reached where the radiation no longer passes through any process liquid. This point corresponds with the level in the vessel.

# CHAPTER 4

# *Measurement of Fluid Flow Rate*

## 4.1 GENERAL

Measurement of the flow rate of fluids through pipes and conduits is one of the important, if not *the* most important, branches of measurement in the process field: there are a great number of instruments on the market for this purpose and more are appearing all the time. As with all industrial measurements, these can be divided into two categories: direct (usually referred to as 'positive displacement') and indirect or inferential. Substandard instruments used for calibration are always of the direct type, as in the case of pressure measurement. A further important distinction must be drawn between the measurement of volume flow rate and mass flow rate, the former being the most commonly required.

Flow-rate measurement is necessary for the purposes of operating plants, but rarely for safety reasons. However, the transfer of process fluids into and out of tankage is often subject to commercial and customs (taxation) constraints, especially at points of delivery to a customer. Consequently, very high accuracy is often demanded of flow rate measuring systems, and calibration is of great importance. It is very important to be clear about the different ways of expressing accuracy which are used by manufacturers for different types of flow meter. In some cases it is expressed as a fraction or percentage of the 'full scale' or maximum flow which can be measured, whilst in others it is expressed as a fraction of the actual flow rate. In all cases the range of flow rates over which the quoted accuracy can be maintained should be given. The difference arises because of the principle of operation of the meter: in some cases the accuracy is a function of features related to the size of the sensing element, whilst in others the error is proportional to the amount

39

of fluid flowing through the meter. A meter operating at 20% of its maximum capacity with an accuracy of ±5% of reading has exactly the same accuracy as another also operating at 20% of its maximum with an accuracy of 1% of full scale.


## 4.2 FLOW MEASUREMENT BY DIFFERENCE OF PRESSURE

Work has to be done on a fluid as it flows through a pipe, both to overcome the internal friction of the fluid (viscosity) as its molecules rub together, and to overcome the friction between the fluid and the pipe walls. However, aside from this gradual loss of energy, the fluid can neither gain nor lose energy as it flows. The total energy of the fluid at any point is made up of:

(a)  kinetic energy (or energy of motion);
(b)  potential energy (or energy of position (height));
(c)  pressure energy.

If a restriction is placed in the pipe, reducing its cross-sectional area, the fluid will have to increase its velocity through this restriction, thus increasing the energy of motion. The energy of position cannot change suddenly, and so a reduction in pressure energy must occur. This phenomenon, which is described by Bernoulli's energy equation, has long been used to measure the rate of flow. A restriction is made in the pipeline, and the difference in pressure measured upstream to that measured at the restriction where the flow velocity is increased has a fixed relationship to the increase in velocity. From the ratio of the cross-sectional area of the pipe to that at the restriction the ratio of the velocity of flow in the unrestricted pipe to that through the restriction can be calculated, and by equating the loss in pressure energy to the gain in velocity energy the flow rate can be calculated.

If the process fluid is a liquid (and therefore incompressible) it can be seen (Fig. 4.1) that the volume flowing past a point in the larger diameter pipe in a given time must be the same as that passing a point in the narrower pipe. Hence

$$A_1 \times V_1 = A_2 \times V_2 \qquad \text{or} \frac{V_1}{V_2} = \frac{A_2}{A_1} \qquad \text{or } V_2 = \frac{A_1}{A_2} \times V_1$$

where $A_1$ and $A_2$ are the cross-sectional areas of the pipe upstream and at

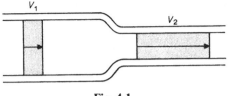

**Fig. 4.1.**

the restriction, respectively, and $V_1$ and $V_2$ are the volumes per unit time flowing past points at these locations in a given time.

The energy of motion (kinetic) is proportional to the square of the velocity of flow; therefore the reduction in pressure when the fluid enters the narrower pipe is proportional to:

$$(V_2)^2 - (V_1)^2 \quad \text{or} \quad (V_1)^2 \times \left[\left[\frac{A_1}{A_2}\right]^2 - 1\right] \quad \text{or} \quad C \times (V_1)^2$$

Hence:

$$(P_1 - P_2) \equiv kC(V_1)^2$$

where $k$ is a constant of proportionality and $C$ is a constant depending on the ratio of cross-sectional area of the pipe upstream and at the restriction.

Naturally it is not practicable to reduce the diameter of the pipe for more than a short length in order to measure flow rate, since the energy absorbed by friction increases in proportion to the square of the velocity. The measuring restriction takes one of the forms shown in Fig. 4.2.

If the fluid is compressible (gas or vapour) it will decrease in density as its pressure falls and its velocity increases in order to flow through the restriction. Thus, a compressibility factor must be included in the relationship between pressure drop and velocity:

$$(P_1 - P_2) \equiv kCY(V_1)^2$$

where $Y$ is the compressibility factor which will be unity for a liquid and greater than unity for any gas or vapour.

The value of the compressibility factor, $Y$, depends on the ratio of the cross-sectional areas of the pipe and the restriction, the ratio of pressures and the ratio of the specific heats at constant pressure and at constant volume (the 'adiabatic' constant). Hence, for any given fluid and any given restriction, $Y$ will be constant, and thus for any fluid and any restriction:

$$(P_1 - P_2) \equiv K(V_1)^2$$

where $K$ is a constant which depends on: (i) the fluid, $Y$; (ii) the ratio of areas, $C$; (iii) the way in which the fluid behaves downstream of the restriction, $k$.

Where the restriction is formed by a simple 'orifice plate' (Fig. 4.2(a)) the diameter of the restriction is not simply that of the orifice, as the fluid, having mass and therefore momentum, cannot change direction suddenly. For this reason $k$ has to be found experimentally for orifice plates, but can be calculated from theory for a properly designed venturi (Fig. 4.2(c)).

As the process fluid reverts to a high pressure state after the measuring restriction there is a tendency for energy-wasteful 'eddies' or 'swirls' to form near any sudden change of shape in the pipe wall. The venturi is designed to minimise this tendency and to ensure that the

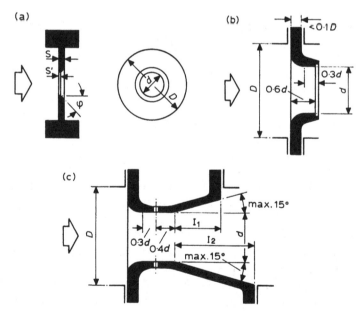

**Fig. 4.2.** (a) Orifice plate. (b) Nozzle. (c) Venturi: $l_1$, short venturi; $l_2$, long venturi.

measurement of flow rate does not impose a large running cost on plant operation due to 'unrecovered' pressure loss; this is often a serious factor where the flows are very large, such as large gas compressor systems. Unfortunately, the large venturi is a very expensive piece of equipment, and often the 'short' venturi is used; these are less efficient in saving energy, but much better than the relatively cheap orifice plate, and cost considerably less than the venturi tube. The angle of the 'expansion' section of the venturi is chosen to enable the pressure in the fluid to 'hold' it against the pipe wall, even though the momentum will tend to make it continue in a straight line and thus leave the wall and form eddies. This does not affect the measurement, and where energy is not so important the 'nozzle' (Fig. 4.2(b)) will give similar accuracy. Both the venturi and the nozzle 'guide' the fluid into the restriction, so that there is no doubt where the low pressure tapping should be located.

The simplest form of restriction used, and by far the cheapest, is the orifice plate; the process fluid is not guided into the restriction and therefore its path is determined, not by the shape of the restricting element, but by the mass and velocity (momentum) of the fluid (see Fig. 4.3). The narrowest cross-section occurs at the 'vena contracta' which is therefore the position at which the pressure differs most from the upstream pressure. The vena contracta occurs at approximately half a pipe diameter downstream of the leading edge of the orifice plate (which must be sharply defined for accuracy) and for maximum sensitivity for a given restriction the pressure tapping points are located

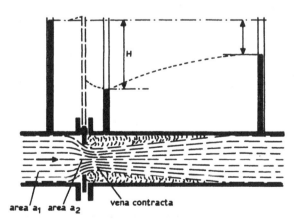

**Fig. 4.3.** Pressure relations for orifice installation.

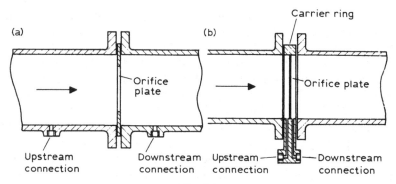

**Fig. 4.4.**

half a diameter downstream and one diameter upstream (the so-called *D* and *D*/2 configuration, Fig. 4.4(a)). Therefore, for a given differential measuring device this enables the energy loss to be minimised when using the cheap orifice plate.

The distance between the pressure tappings is such that these must be made on site, in the pipe wall, which makes it difficult to inspect; the cost of site labour is also high, and these facts have led to the use of 'flange taps' (Fig. 4.5) and 'carrier rings' (Fig. 4.4(b)). In the former case special flanges are supplied for the construction engineer to weld the pipe where orifice plates are to be installed: as these are always of the 'slip-over' type where the weld is made on the outside of the pipe the inspection problem is overcome, and no additional expensive site labour is used to install the orifice plate. Carrier rings are mounted between conventional flanges and are manufactured complete with tappings, isolating valves and a ring (in two halves) in which the orifice plate itself is mounted (allowing for standard manufacture and change of meter range).

The orifice plate itself is normally of the type shown in Fig. 4.2(a) with a 'sharp' edge facing upstream; in fact the accuracy of measurement depends on this edge remaining sharp. However, for relatively viscous fluids accuracy can be preserved at lower flow rates than would be possible with a sharp-edged orifice plate by using a quadrant or 'quarter circle' profile as shown in Fig. 4.5.

The reason for the importance of the edge profile of the orifice plate is that calibration of any restriction-type flow-rate measuring system depends, not upon comparison with a substandard (this type of meter is not usually provided with in-line proving equipment), but upon known

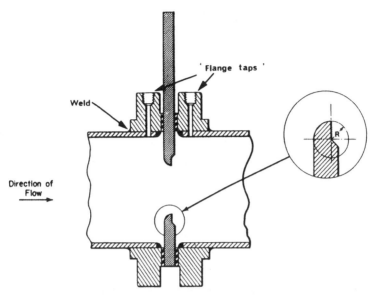

**Fig. 4.5.**

and very well-documented data on the behaviour of fluids flowing in pipes. In particular, the data are only valid for certain values above a minimum of a non-dimensional number, known as the Reynolds number. This number in turn depends on the density, velocity and viscosity of the process fluid.

In special circumstances the shape of the orifice can be other than circular, though the data relating flow rate to the measurement of pressure difference is not so well documented for other shapes; for this and other reasons accuracy will be low. Such a case is shown in Fig. 4.6: an orifice in the shape of a segment of the internal circular shape of the pipe allows dirt to pass freely, whereas a conventional orifice would not.

The accuracy with which flow rate can be measured by pressure difference across a restricting element depends upon the available data for density, compressibility, etc. and upon the precision of measurement of the differential pressure. Obviously, it is important to employ a restriction which is sufficient to provide a readily measured pressure difference over the range of flow rates to be measured; however, a greater restriction than is necessary is very wasteful of energy. If the restriction element is properly designed, accuracy is about ±1% of full scale over a range of flow rates of about 3:1. This low rangeability is the

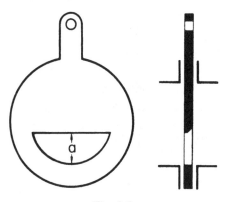

**Fig. 4.6.**

direct result of the square law relationship between flow rate and differential pressure: at 0·3 of maximum flow the pressure is $0·3^2 \simeq 0·1$ of maximum. However, the orifice plate can be replaced by another to change the range if this becomes necessary, and the range can be extended to approximately 10:1 by using three differential pressure measuring devices, each having a measuring range one tenth as large as the next.

In practice the orifice/pipe internal diameter ratio must be less than 0·55 for accurate measurement; at greater ratios (less restriction) the irregularities in the pipe surface seriously affect accuracy. The actual pressure drop will depend upon the velocity of the flow in the pipe, i.e. upon how generously the pipe has been sized for the flow it has to carry. Therefore, if three differential pressure measuring devices are used to measure over a range of 10:1 and the measuring span of the most sensitive is only about 5 cm W.G., the measuring span of the least sensitive will be 500 cm W.G. Using orifice plates, only about one half of this pressure drop will be 'recovered' when the fluid flow cross-sectional area 'expands' downstream of the vena contracta; in other words the pressure loss will be about 250 cm W.G., which represents a considerable energy loss. A much better recovery would be obtained using a nozzle or venturi, but these are expensive elements. Thus the orifice plate, although very cheap (especially in large sizes), is restricted in practice to applications where the range of flow rates to be measured is little greater than 2:1, or where energy loss is unimportant (such as the measurement of the flow rate of gas or crude oil from a producing well, where the natural pressure has to be dissipated).

Nozzles and venturis are very expensive in large sizes, while positive displacement meters are extremely expensive in large sizes, as well as absorbing a lot of energy. There is, therefore, a great incentive to find a less expensive meter, both in terms of capital cost and running cost (energy loss), which has the ability to measure accurately over a range of flow rates of about 10:1. A further grave disadvantage of any flow measurement system using a fixed restriction is that the response is very 'non-linear' because of the square-law relationship between flow velocity and pressure drop. In addition to severely limiting the useful range of the system this means that the scale divisions on any indicator are much closer together at the low flow end of the scale than at the high flow end, making 'discrimination' more difficult. This factor also limits the accuracy of the signal produced by any transmitting device and has a seriously adverse effect when the measurement is used in an automatic control system.

## 4.3 MEASURING GAS FLOW RATES

If the meter is installed above the process line (Fig. 4.7), any liquid which may be carried along by the flowing gas, and which finds its way into the impulse lines can drain back into the process line; this prevents it from collecting in one or other of the impulse lines where it would cause a serious error because of the pressure exerted by a column of liquid. If, however, the meter is installed below the process line, condensate traps of adequate capacity must be installed, as shown in Fig. 4.8.

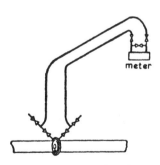

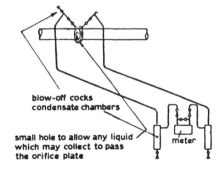

blow-off cocks
condensate chambers

small hole to allow any liquid
which may collect to pass
the orifice plate

meter

**Fig. 4.7.** Installation for clean non-corrosive gases — meter above orifice.

**Fig. 4.8.** Installation for clean non-corrosive gases — meter below orifice.

## 4.4 MEASURING LIQUID FLOW RATES

If the meter is below the process line (Fig. 4.9), vent cocks must be fitted to allow gases and vapours to be released during commissioning and calibration. During normal operation, however, gases and vapours cannot enter the impulse lines as they are lighter than the liquid. If the meter is above the process line, gases and vapours can enter the impulse lines, and suitably sized traps must be installed (Fig. 4.10).

## 4.5 MEASURING VAPOUR (STEAM) FLOW RATES

If the process fluid is a vapour flowing at a temperature higher than ambient, which will condense at ambient, then because there is no flow through the impulse lines condensate will inevitably collect and fill the impulse lines. Provided the head of liquid in each impulse line is exactly equal, no error will be introduced; for this reason the condensate chambers must be exactly level, as shown in Fig. 4.11.

If the meter is below the process line (Fig. 4.12), cocks must be fitted to allow gas, which may have been dissolved in the condensate, to be vented. If the meter is above the process line, gas collecting traps must be added (Fig. 4.13).

## 4.6 MEASURING FLOW RATES IN CORROSIVE, TOXIC OR INFLAMMABLE FLUIDS

If the fluid is corrosive, seals may be used in the impulse lines; however, since the differential pressures generated by properly sized restriction elements are small, their use is not normally recommended. Either a diaphragm meter (see Chapter 2) constructed of suitably resistant materials or a transducer would be used. Seals would not normally be used for toxic or inflammable fluids.

## 4.7 ISOLATING/EQUALISING VALVE MANIFOLDS

In order to maintain, commission or calibrate the flow measurement system, it must be possible to isolate the differential pressure-measuring device from the impulse lines and to equalise the pressure on each side.

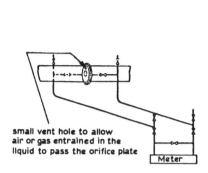

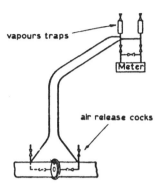

**Fig. 4.9.** Installation for clean liquids — meter below orifice.

**Fig. 4.10.** Installation for clean liquids — meter above orifice (necessary pressure must be available to maintain the liquid head to the meter).

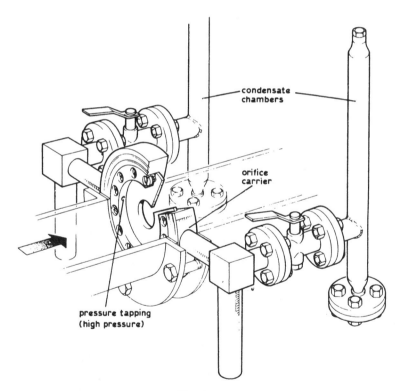

**Fig. 4.11.**

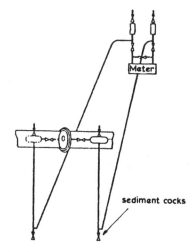

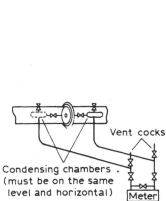

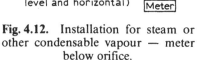

**Fig. 4.12.** Installation for steam or other condensable vapour — meter below orifice.

**Fig. 4.13.** Installation for steam. Pressure of steam in the line must be sufficient to force condensate up to the level of the meter which is above the orifice.

A manifold of valves (capable of tight shut-off), often specially constructed as a single unit, is used (see Fig. 4.14).

There must also be at least one shut-off (isolating) valve in each impulse line at the tapping point. The 'blow-off' valves allow each impulse line to be vented once the isolating valve at the tapping point is closed; the isolating valves then allow the measuring device to be

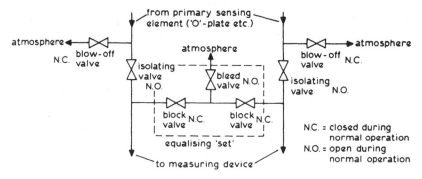

**Fig. 4.14.**

removed. They also allow it to be 'equalised' using the 'block and bleed' equalising set for calibration purposes. In normal operation the two block valves are closed and the bleed valve is open; for equalising this is reversed. If either of the block valves leaks the 'bleed' will be noticed, and flow will not take place in the impulse lines as it would if a single equalising valve were used.

## 4.8 INSTALLATION OF THE PRIMARY SENSING ELEMENT

The location of the orifice plate, etc., in the process line is critical. Bends, reducers or throttling valves will disturb the flow pattern on which the relationship between differential pressure and flow rate depends. In normal circumstances a straight length of pipe of 10 diameters upstream of the primary sensing element will be sufficient to allow such disturbances to subside, but this length depends very much on the cause of the disturbance. Fifty diameters may be insufficient if the cause is an almost closed throttling valve, as may be the case if the process lines are oversized. In such cases straightening vanes must be inserted in the pipeline (Fig. 4.15).

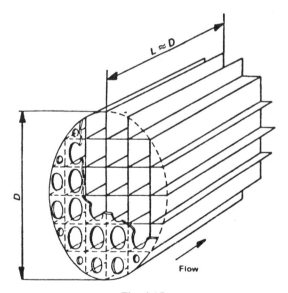

**Fig. 4.15.**

## 4.9 MEASUREMENT OF FLOW RATE BY VARIABLE AREA RESTRICTION METHODS

One solution to the difficulty of 'rangeability', implicit in measuring flow rate by measuring the pressure drop across a restriction, is to vary the cross-sectional area of the restriction rather than the differential pressure. Since the pressure drop is proportional to the ratio of the cross-sectional area of the restriction to that of the pipe, no square root is involved and the rangeability of flow measurement is not restricted as it is when using measurement of differential pressure across a fixed restriction. A simple 'variable restriction' measuring system is shown in Fig. 4.16(a).

A servo mechanism could be added to this system so that the 'gate' is driven to a position such that the differential pressure remains constant whenever flow rate changes. The disadvantage of this method is that the shape of the orifice, as well as its cross-sectional area, changes and in consequence it is not possible to obtain a reliable set of data on which to base an accurate calibration of the system. Such systems cannot be used for accurate flow measurement.

The principle of variable area restriction is used extensively in the orifice and plug type of meter (Fig. 4.16(b)). Just as the positive displacement meter is in effect a pump, which, instead of driving the process fluid, is driven by it, so the orifice and plug meter is like a control valve, which, instead of regulating the flow by reason of the relative position of plug and orifice, is positioned by the flow.

One very big advantage of the plug and orifice type of meter is that it is

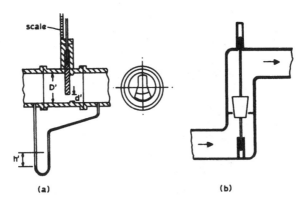

**Fig. 4.16.**  (a) Gate-type area meter; (b) orifice and plug meter.

tolerant of dirt; the plug will move to open the orifice and allow dirt to pass through.

Modern versions of the instrument use a conical-shaped glass tube as the orifice, so that the 'plug' (float) moves up and down the whole length of the tube as flow rate changes from minimum to maximum (Figs 4.17 and 4.18). A magnetic follower can be used to operate an indicator pointer or transmitter mechanism.

This type of flow meter is capable of measuring to accuracies of about ±1% of full scale over a range of 10:1. Calibration depends entirely upon the construction of the glass tube which is made by special manufacturing processes, and since the measuring tube has to be the same diameter as the process line they are expensive meters.

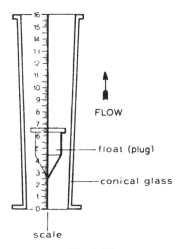

**Fig. 4.17.**

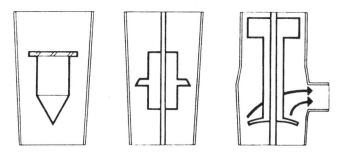

**Fig. 4.18.** Various shapes of float.

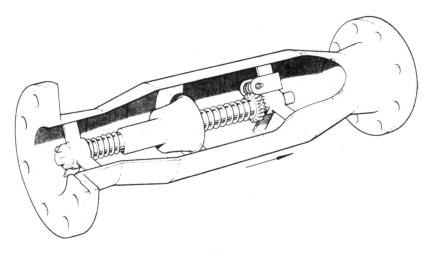

**Fig. 4.19.**

The principles of differential pressure measurement and variable area measurement have been combined in the 'Gilflo' type of meter shown in Fig. 4.19. The plug or cone takes up a position in the orifice which is determined by the balance of forces generated by the spring on the one hand, and the differential pressure acting on the cone on the other. The position and shape of the cone together determine the size of the orifice; the shape of the plug is so arranged that the differential pressure generated is linearly related to flow rate and not, as in the case of a fixed restriction, to the square of flow rate. In effect the size of the orifice is varied to suit the flow rate; as a result the useful range of the meter is greatly extended.

The manufacturers claim a remarkable 100:1 range with a repeatability of $\pm\frac{1}{4}$% of maximum reading; however, this is to some extent of academic interest only, as at 1/100 of maximum flow rate the repeatability expressed as a fraction of actual flow rate is then $\pm25$%. At the lower limit of the range at which a turbine or positive displacement meter can reliably be used — 20:1 (i.e. 5% of maximum) — the repeatability of the 'Gilflo' would be $\pm5$%, which does not compare favourably; however, at even lower flow rates it would increasingly provide a better performance. In fact a turbine meter cannot be used at all at a flow as low as 1/100 of maximum, and it is doubtful if a positive displacement meter would operate at this sort of 'turndown'. This is a good example of the influence of the principle of operation on the performance of a flow meter; the 'Gilflo' is obviously a suitable meter to use where wide 'rangeability' is the most important requirement, but it would not be suitable for custody transfer applications.

The makers claim that this type of meter is insensitive to 'upstream conditions' (see Section 4.8) which represents a further advantage.

## 4.10 POSITIVE DISPLACEMENT METERS

This type of meter operates on the same principle as the vane, piston or gear pump, with the difference that the meter is driven by the flow of fluid whereas the pump drives the fluid. A fixed *volume* of fluid is transferred from the inlet to the outlet for a given angle of rotation of the impeller shaft. Provided there is no leakage of the fluid past the sealing surfaces of the vanes, there is an exact relationship between the rotational velocity of the shaft and the fluid flow rate. A small amount of energy is absorbed in driving the rotor and this is provided by pressure difference between inlet and outlet. The same pressure difference will cause leakage across the seals if these are imperfect: hence the performance of a positive displacement (PD) meter depends on good seals and on minimising the friction losses in driving the meter. Gear-type PD meters are typified by gears, which are cut to provide a rolling rather than sliding action which is less likely to cause leaks. In addition to the sliding or rolling seals of the vanes or gears there must be a good seal at the ends of the impeller. There are many types of PD meter on the market, but all operate on the same principles: they can be built to stand very great pressure and are capable of staying within very accurate calibration for long periods of service when used to measure flows of viscous fluids (oils) which serve to lubricate the seals. Moreover, they are the only type of meter which actually becomes more accurate as the viscosity of the fluid increases.

In the 'rotary piston' type of PD meter shown in Fig. 4.20 the central journal of the rotary piston travels around in the groove; the rotary piston itself is not free to rotate as the 'separating wall' slots through it. It therefore oscillates in the fashion which can be seen in Fig. 4.20 (i) to (iv) and by doing so transfers a volume of fluid equal to the capacity of volumes 1 and 2 from the inlet to the outlet port of the meter. This is accomplished as follows. In position (i) fluid enters both volume 1 and volume 2 spaces under the high upstream pressure, whilst fluid is still leaving volume 1 through the outlet port. High pressure acting on the

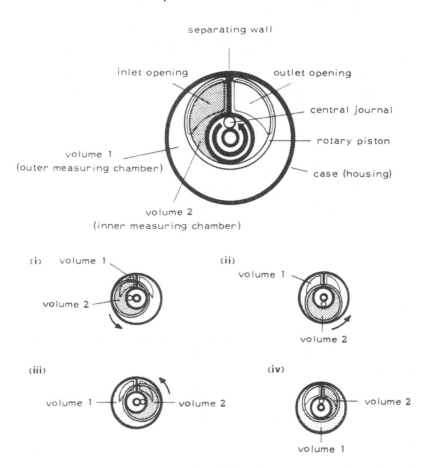

**Fig. 4.20.**

separating wall and piston causes it to rotate. In position (ii) volume 2 space is full of liquid which has just entered from the high pressure upstream, and the inlet and outlet ports are both closed. In position (iii) fluid from volume 2 is leaving through the outlet port and has started to enter the rotary piston on the other side of the separating wall. High pressure on the piston and separating wall continues to cause rotation of the piston. In position (iv) volume 1 is full and about to discharge.

It should be noticed that the rotary piston does not in fact rotate, though its central journal does; the piston itself moves from side to side and up and down, being constrained by the separating wall which separates the high upstream pressure from the lower downstream pressure. In one revolution the meter transfers a volume of liquid equal to the combined volumes 1 and 2 from the inlet port to the outlet.

Another type of positive displacement meter known as the oval wheel meter is shown in Fig. 4.21. The distance between the spindles of the two 'oval wheels' is fixed, as is the sum of the major and minor axes of the oval (both 'wheels' are identical). They are geared to rotate together as shown in Fig. 4.21 (i) to (iii). In position it can be seen that a volume of fluid which has entered from the upstream high pressure has been trapped and will be delivered to the discharge port as the rotation continues. At all times the high pressure is sealed from the low pressure side and so this volume is delivered for each revolution of the meter, provided there is no leakage past the seals.

In all PD meters the mechanism should be as free as possible to rotate so that the difference in pressure between upstream and downstream is as small as possible, thus minimising leakage which will be proportional to pressure difference. For the same difference in pressure, increased viscosity actually decreases leakage error.

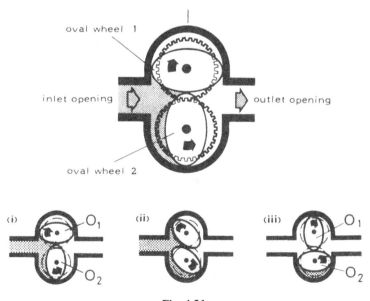

**Fig. 4.21.**

## 4.11 TURBINE METERS AND IN-LINE PROVING

Turbine meters differ from PD meters in that there is no positive seal between the high upstream pressure and the lower downstream pressure. Like PD meters they are driven by this pressure difference, but unlike them any increase in the power required to drive the impeller will result in a change in calibration rather than an increase in pressure drop; therefore, any increased bearing friction, etc., affects accuracy in a way it does not in the case of the PD meter.

Turbine meters such as those shown in Fig. 4.22 have long been used to measure *and totalise* (see Chapter 5) the flow of utilities fluids — fuel

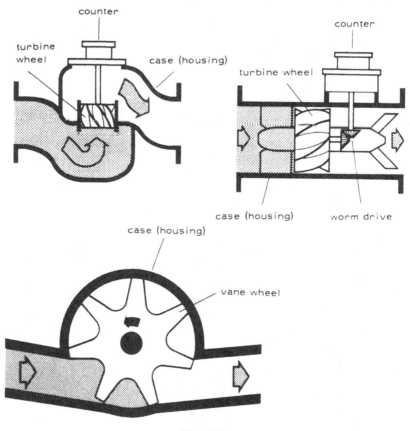

**Fig. 4.22.**

oil, water, steam, etc. The impeller rotates at a speed which is proportional to the flow rate of the fluid, and it is easy to count the total number of revolutions in order to obtain a totalised flow measurement; thus, the rate of revolution provides a measure of the flow rate. The relationship is varied by any change in fluid viscosity, so that accuracy of these meters depends on the fluid viscosity remaining constant.

Positive displacement meters have been used for a long time in applications where very high accuracy is essential, e.g. custody transfer. In large sizes, however, they are very large and very expensive, and with the invention of the in-line prover, high precision turbine meters (Fig. 4.23) have taken their place, except for the measurement of high viscosity liquids. An impeller with blading similar to a gas or steam turbine is driven by the flow of fluid. The bearings of the impeller are almost frictionless, and the speed of rotation of the impeller is measured by electronic counting of the pulses generated by a small magnet embedded in the impeller as it passes the 'pick-up' coil. The energy absorbed by the metering element is very small, response is linear and with in-line proving equipment accuracy is comparable with that of a positive displacement meter over a 10:1 or greater range of flows. These meters are very much cheaper than positive displacement meters, especially in the larger sizes. When measuring clean fluids they compare favourably with venturi nozzles with regard to price and are more accurate over a much wider range. Performance deteriorates as

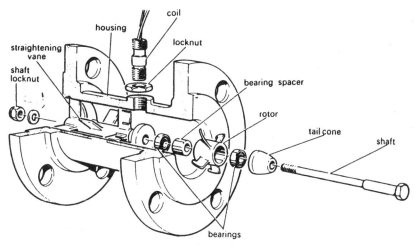

**Fig. 4.23.** Component parts of a turbine flow meter.

viscosity of the process fluid increases; they cannot operate in dirty fluids as the bearings seize up. By their nature they operate electronically and must be certified safe to use in areas of the plant where there is fire/ explosion hazard.

The installation of an in-line prover in addition to a flow meter makes it possible to check the calibration of the meter *whilst it is operating* as often as necessary, and with very little trouble. The greater susceptibility of the turbine meter to changes of calibration (as compared to the PD meter) is not an important factor, therefore.

The meter prover is not itself a measuring device, but operates on the basis of stopping and starting a special counter, a sphere, which is a close fit in the pipe and travels along with the process fluid, sweeping out an accurately known volume (Fig. 4.24). The counter counts the revolutions of the meter impeller and is started and stopped by 'detector' switches at the start and finish of the calibration run. The sphere stops in a section of pipe which is specially enlarged so that the fluid can flow past it: another calibration can be made by simply changing the position of the 4-way diverter valve so that the direction of flow through the calibrated section is reversed, carrying with it the sphere. The prover is itself calibrated by a special calibrating service and, provided that it is not damaged, will not change over very long periods.

## 4.12 MAGNETIC FLOW METERS

This type of meter uses the process liquid as an electrical conductor, which by flowing through a magnetic field has an electric current induced in it. It consists of an electromagnet wrapped around a length of process pipe, which is lined with an insulating material as shown in Fig. 4.25: electrodes are installed in the wall of the pipe on opposite sides, and these enable an electrical circuit to be formed through the liquid and the measuring device (a galvanometer). This type of meter imposes no energy loss as there is absolutely no obstruction to flow: it is very suitable for measurement in corrosive, abrasive or very dirty liquids. It cannot be used to measure the flow rate of gases or low-conductivity liquids (which unfortunately includes hydrocarbons) but is very suitable for most water applications. It also has the considerable advantage over most other flow meters that it is totally insensitive to flow profiles, turbulence, density or viscosity, as it measures the average velocity in the pipe and thus the true volume flow rate. If the process

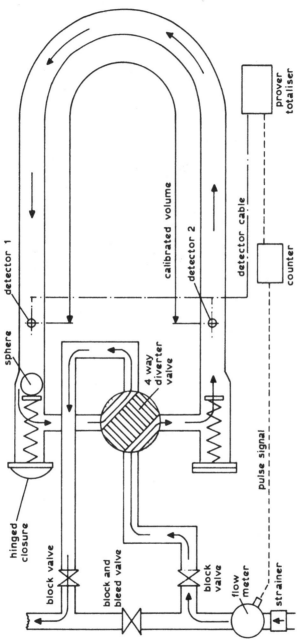

**Fig. 4.24.** Diagram of bidirectional meter prover.

62

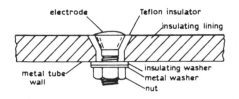

form of electrode

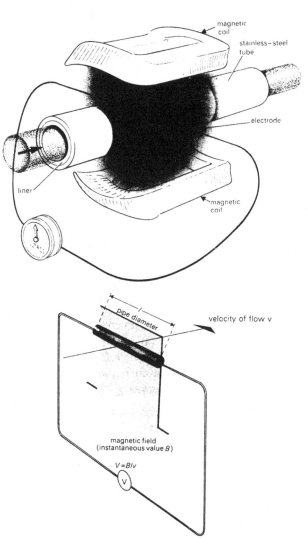

Fig. 4.25.

liquid deposits a film on the wall of the pipe which is significantly less conductive than the liquid itself, the accuracy will be affected as the electrodes/liquid resistance will increase. The range of the instrument has a lower limit because the velocity of the fluid must be great enough to cause a measurable electric current to flow, but it has no upper limit except that imposed by the current-measuring instrument and the practical limits of flow rate in the process pipe (imposed by pumping pressure and frictional resistance). Accuracy of measurement is therefore largely a function of the electronic measurement of the electric current and of the magnetics of the meter; it is usually quoted by the makers as ±1% of the maximum flow rate recommended for the instrument.

## 4.13 ULTRASONIC FLOW METERS

Energy can be transmitted through liquid by alternating pressure waves in the same way that electrical energy travels through an electrical conductor; both travel at an exactly constant speed, though pressure waves travel many times slower than electricity. These facts are used to measure flow rates in liquids as shown in Fig. 4.26.

For energy to travel from transmitters/receivers A to B it must cover a distance ab; however, if the fluid is moving, the distance travelled

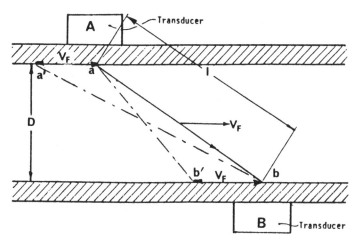

**Fig. 4.26.**

through the fluid is ab' which is shorter. Energy travelling from B to A must also travel distance ab, or rather ba in this case, but since it travels against the direction of flow, the distance through the moving fluid is ba', which is longer. The transmitters/receivers A and B take it in turn to transmit short bursts of energy through the liquid, at the same time signalling the other transmitter/receiver by an electrical signal (which travels many times faster) that it has done so. The 'time of flight' for ab' is compared with that for ba', the difference being proportional to the flow rate.

If the fluid contains a lot of solid matter, the pressure energy is scattered and little is received at the receiver. In such cases it is better to use the type of meter shown in Fig. 4.27. This type, called a 'Doppler shift' meter measures the 'phase shift' of signals received directly at B as compared with signals received by reflection (Fig. 4.28) (note that the beam has to be narrow). This phase shift is proportional to flow rate. This type of ultrasonic meter only works when there is plenty of dirt or air bubbles in the fluid to cause reflection.

This type of meter offers no obstruction to flow, has no lower limit of flow beyond which it will not operate (unlike the magnetic flow meter) and will operate on almost any liquid. The makers claim rangeability of up to 200:1 and repeatability of ±0·1% of maximum flow rate, but only ±2% accuracy because this must depend on factory calibration (unless in-line proving equipment is installed). The cost does not depend on pipe size.

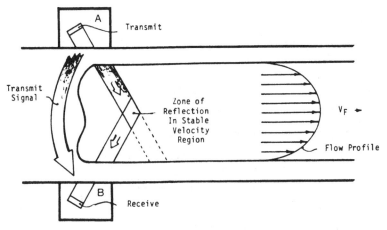

**Fig. 4.27.**

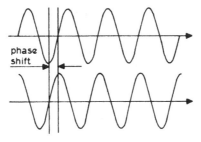

**Fig. 4.28.**

## 4.14 'VORTEX' METERS

This type of device uses a specially shaped obstruction, located in the centre of the pipe, to cause the formation of vortices in the same way that a protruding rock will sometimes do in a fast-flowing stream. It has been found that when 'vortex shedding' occurs, the frequency with which vortices are 'shed' is extremely accurately related to the flow rate over a very wide range. The makers claim accuracy of ±1% of maximum rated flow over a range of up to 100:1. The shedding of a vortex is detected by a very sensitive temperature or pressure detector set into the obstruction or by a sonar-type transmitter/receiver as shown in Fig. 4.29. The detectors set into the obstruction detect slight differences in pressure and temperature which occur as a vortex is 'shed' first from one side and then from the other of the obstruction. The sonar type of detector relies

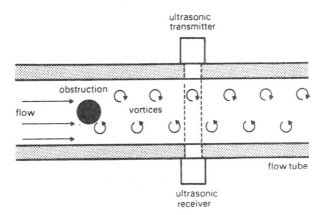

**Fig. 4.29.** Vortex flow meter.

on the signal at the receiver being disturbed as each vortex passes it. In both cases the pulses generated are counted by electronic circuits and the lower limit of the instrument range is set by the number of pulses generated by each unit of volume flow.

## 4.15 THE FLUIDIC FLOW METER

This type of meter is rather similar to the vortex meter in that it utilises the slight temperature difference which occurs when the flow of fluid across a sensitive thermistor element ceases. The principle of operation depends upon the 'coanda' effect by reason of which a moving column of liquid, when faced with diverging pipe walls, will cling to one in preference to the other. As can be seen from Fig. 4.30, part of the flow is diverted through a passage on this side and serves to divert the main flow away from this wall, so that it then clings to the other wall; this action occurs repeatedly and the main stream oscillates rapidly from wall to wall. The frequency of oscillation is linearly proportional to the

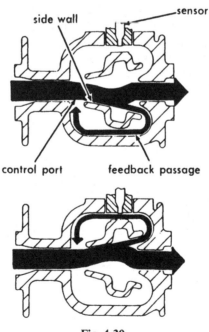

**Fig. 4.30.**

flow rate, accuracy being about ±1% over a range of flows which depends on the viscosity of the liquid, but may be as little as 3:1 or as much as 100:1. Detection of the flow/no flow change in the two bypasses is much more certain than with the vortex meter.

## 4.16 MASS FLOW MEASUREMENT

All the meters described so far measure the volume flow rate directly (positive displacement) or infer it from measurement of velocity. The mass flow rate is sometimes required, and can be obtained by computation from measurement of both volume flow rate and density. This is complex and expensive; however, for flow rates up to about 1000 kg h$^{-1}$ of liquid it is possible to measure mass flow rate directly using the meter shown diagrammatically in Fig. 4.31.

This meter comprises a positive displacement pump and four carefully matched orifice plates, arranged in the form of a 'bridge' as shown. The differential pressure, measured across the pump, is linearly proportional to the mass flow rate $Q$. This is shown below.

Since the orifices are identical, the velocity of flow through orifices a and d are given by

$$(P_1 - P_2) = kC(V_a)^2 \times \text{density}$$

and

$$(P_1 - P_3) = kC(V_d)^2 \times \text{density}$$

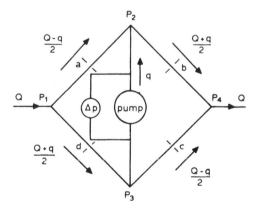

**Fig. 4.31.**

and since the pipe sizes are identical

$$(P_1 - P_2) = K\frac{(Q - q)^2}{4} \times \text{density}$$

and

$$(P_1 - P_3) = K\frac{(Q + q)^2}{4} \times \text{density}$$

or

$$\frac{(P_1 - P_2)}{\text{density}} = \frac{K(Q^2 - 2Qq + q^2)}{4}$$

and

$$\frac{(P_1 - P_3)}{\text{density}} = \frac{K(Q^2 + 2Qq + q^2)}{4}$$

By subtracting these equations we obtain

$$\frac{(P_1 - P_3) - (P_1 - P_2)}{\text{density}} = KQq$$

or

$$(P_2 - P_3) = KQq \times \text{density}$$

and since $q$ is constant (positive displacement pump)

$$(P_2 - P_3) \propto Q \times \text{density (which is mass flow)}$$

Accuracy for this type of meter is quoted as $\pm(0.5\%$ of actual flow rate $+0.02\%$ of maximum flow rate) over a range of 30:1.


## 4.17 STREAMLINE/TURBULENT FLOW AND VELOCITY PROFILE

The 'pattern' of flow of the fluid in the pipeline is very important and must be known when using most of the inferential methods (the only real exception being the magnetic flow meter). Depending upon the velocity and viscosity, two very different patterns are possible — streamline or turbulent. Streamline flow takes place at low velocity and high viscosity, and is, as the name implies, a pattern in which each

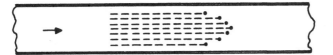

**Fig. 4.32.** Streamline flow.

'filament' of the flow travels in a straight line parallel to the axis of the pipe (Fig. 4.32). The pattern becomes 'turbulent' when the velocity increases to the point where these straight line 'filaments' break up into eddies and swirls. These eddies and swirls are initiated by small obstructions in the pipe, notably roughness on the pipe wall. If flow rate is increased smoothly in a pipe which is exceptionally smooth and without the normal obstructions found in process pipework, the onset of turbulent flow can be delayed, but normally it is a function of velocity and viscosity. The distribution of the velocity of flow across the pipe under streamline flow is as shown in Fig. 4.33(a) — almost stationary near the wall of the pipe and gradually increasing towards the centre. Such a pattern makes measurement of flow rate very difficult, for it will change (as shown by the dotted forms in Fig. 4.33(a)) as the flow rate increases or decreases, making the generation of differential pressure across an obstruction or the speed of rotation of a turbine a function not only of the average flow velocity but also of the velocity distribution. For this reason flow rate measurement by these means is only possible when the flow pattern is *fully* turbulent, when the velocity distribution will be as shown in Fig. 4.33(b), more or less constant across the pipe diameter with only a very small 'transition' area close to the pipe wall. Even so most of the errors of measurement are associated with this region, which

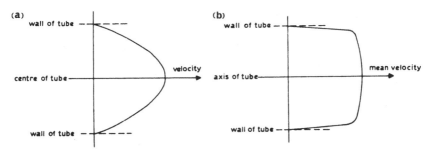

**Fig. 4.33.** (a) Streamline flow velocity distribution; (b) turbulent flow velocity distribution.

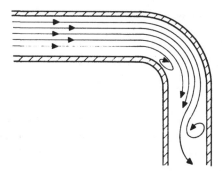

**Fig. 4.34.**   Flow lines in a bent section of pipe.

explains the reduction of accuracy at low flows using orifices and nozzles.

With certain exceptions, all inferential meters are liable to be adversely affected by anything upstream of the meter which will alter the normal flow of the fluid in the pipe. Anything which will impart a swirling motion to the fluid is to be particularly avoided, such as a sharp bend (Fig. 4.34) or a partly open valve (flow meters must *never* be located downstream of a control valve). Any sudden enlargement of the process line will also affect the flow pattern and cause errors in measurement. Orifice plates, nozzles and turbine meters are most seriously affected, whilst magnetic flow meters and variable area orifice meters are least affected (positive displacement meters are not affected at all). Minimum lengths of straight unobstructed pipe of constant diameter which must be provided in the piping design of the plant in order to avoid errors arising in this way are recommended in any standard on flow-rate measurement (Fig. 4.35): these lengths vary from 10 diameters for a single 90° bend to 100 for a $\frac{3}{4}$-closed globe valve (most control valves are of the globe pattern). Where the problem is swirl, as in the case of two 90° bends in different planes, these lengths can be reduced by fitting into the process line a 'flow straightener' (which divides the pipe into a number of small passages over an appreciable length); in other cases this will not help.

These requirements may be an important factor in choosing the most appropriate meter, particularly in the case of retrofitting, since it is often very difficult to accommodate these requirements into the piping design.

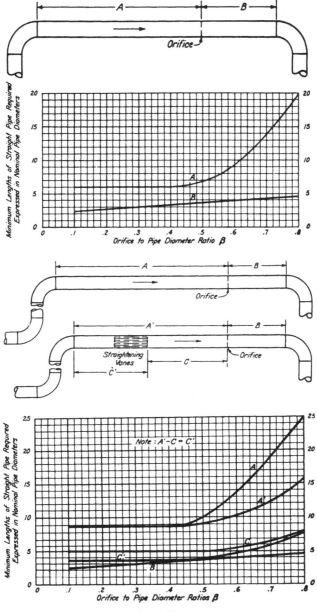

**Fig. 4.35.**

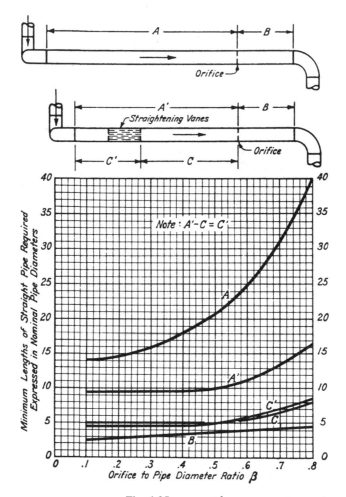

**Fig. 4.35.** —*contd.*

## 4.18  SELECTING A FLOW METER

There are many considerations which determine the selection of a flow
meter for a particular application:

- accuracy and range required;
- nature of fluid: dirty or clean, corrosive, gas, vapour or liquid,
  temperature, pressure, conductivity, viscosity;

- velocity of flow (pipe sizing);
- restrictions on piping layout;
- explosion and fire risk;
- linearity;
- energy conservation;
- mass or volume flow rate;
- cost and maintenance.

The accuracy of measurement of any of the meters described in this unit is better than ±1% of the maximum flow to be measured when the flow rate is actually greater than one third of this maximum. Positive displacement and turbine meters will provide much greater accuracy over a wider range, but are expensive; the Gilflo, fluidic and vortex types are not so accurate but all give very wide rangeability. Orifice plate meters are simple, relatively cheap and rugged, and can be changed to provide a different range, but are restricted in range and accuracy. Magnetic and sonar types offer no restriction to flow and have no real upper limit; however, both require a certain lower limit of flow rate, which in practice may severely limit their range for some applications (depending on how generously the process line has been sized). On the other hand these types, and to a lesser extent nozzles and venturis, are not wasteful of energy. Magnetic flow meters cannot be used on petroleum fluids which are non-conductive electrically, and careful consideration must be given before using any of the electrical output types where there is a fire or explosion risk. Orifice and turbine types are particularly prone to error due to piping-generated disturbances and viscosity changes in the fluid, as, to a lesser extent, are most other meters.

These are only a few of the comparisons that can be made and no firm rules can be given for the choice of meter.

## 4.19 INSTALLATION, MAINTENANCE AND CALIBRATION

Great care must be taken with the installation of any flow meter to ensure that nothing is done which may impair its operation; in this respect it is obviously essential to understand the principle of its operation.

Positive displacement meters depend for continued accuracy entirely on the maintenance of their sealing surfaces; it is obviously important, therefore, that they should never be installed without adequate upstream filters and strainers. What is adequate is a matter for decision by an engineer for each particular case, but it is worth remembering that it may be best to seek another type of meter if the flowing fluid contains very fine abrasive material. Turbine meters, too, will be damaged by fine abrasive material which can get into their bearings, and must be protected (by a strainer) from large solid particles which could damage their rotor blades. The energy loss that filters and strainers impose must be allowed for when the pumps and pipes are sized at the design stage; they must be cleaned regularly in use to prevent the pressure loss increasing to the point where it affects process operation.

Orifice plates, nozzles and venturis, and variable restriction devices are not capable of the accuracy of PD meters and turbines, but are relatively tolerant of dirty fluids. Vortex, magnetic and fluidic types are also relatively tolerant of dirt but sonar types may suffer from signal 'scattering'. However, scaling or corrosive fluids may necessitate the use of 'purges' on *both* low and high pressure connections to any form of differential pressure measuring system. All these types are to some degree affected by upstream conditions, except the turbine meter and also the magnetic flow meter which cannot be used on petroleum fluids.

Many of the more recent types of meter rely on electrical/electronic detection techniques (i.e. sonar, fluidic, some vortex, etc.) and it is essential to ensure not only that they *can* satisfy relevant electrical requirements and standards, but also that they are so installed that they *do* in fact satisfy them.

Flow meters need to be maintained and to have their calibration checked from time to time; it is necessary to ensure, therefore, that the position in which they are installed is such that they are readily accessible.

There are only two ways in which a flow meter can be calibrated; in line or out of line. Calibration out of the process line is almost invariably impracticable on plant, and necessitates sending the meter back to the manufacturer or another calibration authority. For most applications this is not practicable and this is one of the main advantages of the differential pressure measurement systems. Provided an orifice plate is not physically damaged by some large particle carried with the flow, accuracy of measurement depends only upon the differential pressure measurement device which can readily be removed for calibration. Positive displacement meters depend for accuracy on their seals, and if adequately protected by filters will not need calibration unless severely damaged; for this reason they are sometimes used as substandard meters by providing a 'spool piece' in the process line which can be temporarily replaced by a PD meter in order that it can be used to calibrate the meter permanently installed. For highly accurate measurement purposes, PD or turbine meters are usually installed with in-line proving equipment; however, for custody transfer applications on large gas pipelines, nozzles or even orifice plates are sometimes used with three differential pressure measuring devices. With very careful installation much better than the usual ±1% accuracy can be obtained, whilst computers can be used to linearise the signal. This is because of the enormous cost of PD and turbine meters in large sizes. The calibration of most other types of meter cannot be checked in-line, and they are usually accepted as remaining for very long periods within the manufacturer's specification, which is usually conservative. It can be seen from the meters described in this section that the principles of operation are always such that there is little possibility of the calibration changing unless the meter is damaged. If this were not true the meter would not normally find acceptance for industrial use.

## 4.20 NON-FULL-FLOW MEASUREMENT

It has so far been assumed that the size and shape of the pipe or duct is such that a 'full flow' meter can be installed, i.e. a meter that measures the total flow through the pipe or duct. However, this is not always the case, for instance, where the flow of fluid will not permit any obstruction (e.g. the overheads on a distillation column) or where the pipe or duct is simply too large (e.g. heating or ventilating ducts or large steam mains).

If an obstruction can be tolerated, shunt metering may provide an answer (see Fig. 4.36). If no obstruction can be tolerated, the *velocity* of flow can be measured using a pitot tube (Fig. 4.37) and the velocity

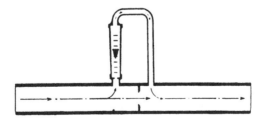

**Fig. 4.36.** By-pass variable aperture meter.

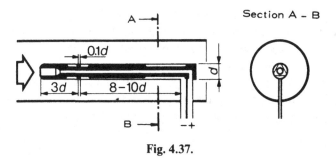

**Fig. 4.37.**

profile either assumed (for an approximate measurement) or determined by a search technique involving moving the position of the pitot in the duct or pipe cross-section. The pitot tube converts the energy of motion to pressure head at the tip and at the same time measures the 'static' pressure in the pipe or duct. The difference in these two readings is proportional to the square of the fluid velocity at the point of measurement.

# CHAPTER 5

# *Measurement of Quantity*

## 5.1 GENERAL

The quantity of material (volume or mass) produced, or transferred into or out of storage, is a very important function of measurement technology. It involves the measurement of the level in storage tankage, the measurement of the weight of vessels or containers with their contents, and the measurement of *total* flow quantities as opposed to flow *rate*. This chapter is concerned with the way that the techniques of measurement of flow rate, level and weight, which have been developed in earlier chapters, are adapted for the purposes of quantity measurement.

Measurements of quantity are required for batch processing, custody transfer and to satisfy customs and excise requirements. In the last two cases the accuracy required is often an order higher than is usual in normal process operations ($\pm 0.1\%$ of actual measured value).

## 5.2 THE PRINCIPLES OF MEASURING TOTAL FLOW

The total flow through a line is the sum of all the fluid which passes through the flow meter over a given period of time. If the flow meter measures the *rate* of flow at any instant, and if the rate does not vary over the given period, the total flow is given by the product of flow rate and time:

$$\text{total flow} = \text{flow rate} \times \text{time period}$$

If, however, the rate varies significantly over the period, this relationship no longer applies. Nor can it be assumed that the correct answer will be

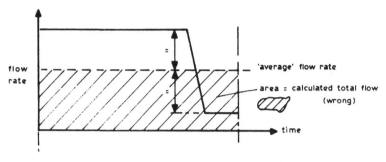

**Fig. 5.1.**

obtained if the 'average' flow rate is multiplied by the time period, as can be seen from Fig. 5.1.

The true answer is given by the 'weighted average' or mean flow rate (over the period) multiplied by the time period (Fig. 5.2(a)). It is difficult to find the 'mean' flow rate over a period if the flow rate varies randomly (Fig. 5.2(b)). In any case the calculation of total flow by this method

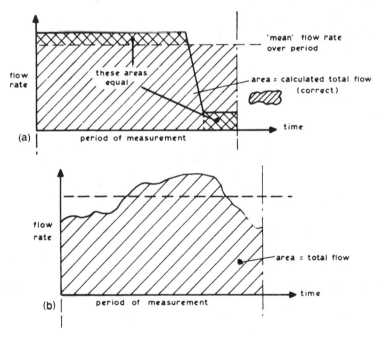

**Fig. 5.2.**

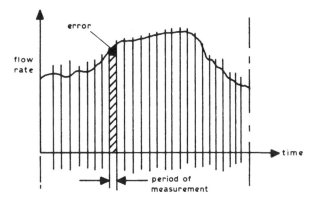

**Fig. 5.3.**

involves recording the flow rate over the whole period of measurement and calculating the total flow later, which is neither convenient nor practical. What is required is a method which automatically 'updates' the total flow from the time that measurement began to the present.

This is achieved in practice by assuming the flow rate is constant over a very short period of time; the 'totalising' mechanism multiplies the flow rate measured by the period and adds this quantity to the previous total (Fig. 5.3). It can be seen that the period must be small enough for the flow rate not to change significantly over the period, if the error, which will be the sum of the errors in each period, is not to be considerable. If the period is very short indeed, it appears that this process of *integration* is continuous, and under these conditions the errors are very small.

All 'inferential' flow meters measure rate of flow, and a mechanism must be added to them which 'integrates' this rate, in order that total flow can be measured. On the other hand, direct or non-inferential meters (i.e. the positive displacement type of meter) do in fact measure true total flow — *not* flow rate. A positive displacement meter functions by transferring a known volume of fluid (determined by the volume of the measuring chamber of the meter) from the inlet to the outlet. Provided that the seals in the meter do not leak, there is no error: the period of measurement is the time taken for the measuring chamber to travel between inlet and outlet (usually the time of one revolution of the meter signal output shaft). A graphical representation of the flow rate would look like that shown in Fig. 5.4: the volume flowing in one

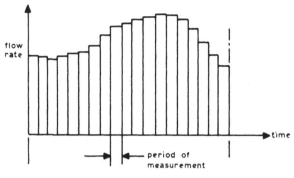

**Fig. 5.4.**

measurement period is known *exactly* and there is *no* error at all in the integration process.

A positive displacement meter is, therefore, the ideal meter for total flow measurement, since it is only necessary to count the revolutions of the output shaft to obtain *exact* integration of the flow rate. All other flow meters are inferential and therefore measure flow rate: an integrating mechanism must be added to make them into a total flow meter, and often this mechanism adds to the errors of measurement.

## 5.3 THE PRINCIPLES OF MEASURING TOTAL VOLUME

Measurement of the total volume of fluid resident in a tank or vessel is just as important as measurement of total flow in or out for the purposes of process operation and inventory, and for customs requirements. The volume can be calculated, in the case of a vessel with constant cross-sectional area, as the product of the level (measured relative to the bottom of the vessel as datum) and cross-sectional area (Fig. 5.5).

However, if the cross-sectional area varies, as for instance in the case of a TEL storage sphere, the total volume in the vessel depends not only on the level, but also on the cross-sectional area; it is in fact the integral of the product of the cross-sectional area at every level and the level, just as the total flow is the integral of the flow rate at different times and the time (Fig. 5.6).

The larger the cross-sectional area of the vessel or tank, the greater the error will be owing to variation of the cross-sectional area, and accurate measurement of the volume contained, in large petroleum storage tanks

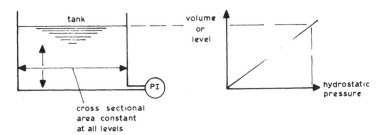

**Fig. 5.5.**

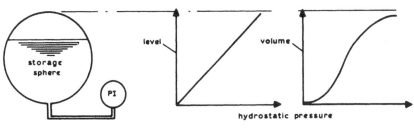

**Fig. 5.6.**

for instance, demands not only accurate measurement of level but also accurate calibration based on measurements of the cross-sectional area at different levels. These measurements are not easy to make when the tank is full of product: however, the stress developed in the wall of such large diameter tanks by the weight of fluid is such that it is forced into a truly circular form at all levels (as the circular form maximises the volume) and, since the ratio of cross-sectional area to circumference of a circle is always the same, the measurement can be made by measuring the circumference of the tank at different levels, which is known as 'strapping'.

## 5.4 THE PRINCIPLES OF MASS MEASUREMENT

The quantity of material for sales purposes can be defined as a volume, or as a mass. In the past it has been found to be too difficult to measure in terms of mass and it has become traditional to sell by volume measure: petroleum is sold by volume even though this is illogical as its energy value is related to mass. Measurements of total flow or total volume can be converted into mass by correcting for density variation, but for accurate measurement the added source of error (in addition to level measurement errors and cross-sectional area errors) makes the final measurement too unreliable. The same problems apply to total flow measurement, since there are no positive displacement mass flow meters and the only inferential types which can measure mass flow rate are either very complex and expensive or make use of the principle of correction for density variation.

The development of the load cell makes it possible to weigh a vessel accurately, but there are considerable engineering problems involved in suspending large tanks for weighing. The weight of the tank itself in relation to its contents is often so great that the errors made in weighing tank and contents are large when compared with the weight of the contents alone. In the case of large petroleum storage tanks, the application of very accurate measurement of hydrostatic head combined with corrections for 'strapping' to provide direct measurement of mass is a possibility.

## 5.5 MEASUREMENT OF TOTAL FLOW USING INFERENTIAL FLOW RATE MEASUREMENT

### Linear Flow Meters

Provided the impulse (signal) produced by the meter is proportional to the flow rate of the fluid (as in the case of variable area meters, turbine meters, fluidic, ultrasonic and magnetic flow meters), it is relatively easy to design a mechanism which will continuously multiply the impulse by time and thus give a signal output proportional to total flow. In the case of the turbine meter this is done by counting the pulses generated when the magnet embedded in the rotor passes the pick-up coil, instead of measuring the *rate* of pulse generation (which is proportional to flow rate).

### Non-linear Flow Meters

Measurement of total flow using meters which produce a differential pressure across a restriction device is not so easy. The impulse signal is proportional to the square of flow rate, and multiplying this by the time will not give a signal proportional to total flow. The impulse signal must first be 'processed' to produce a signal which is proportional to flow. This processing is sometimes carried out by the pressure-measuring device shown in Fig. 5.7; the manometer has one 'leg' specially shaped,

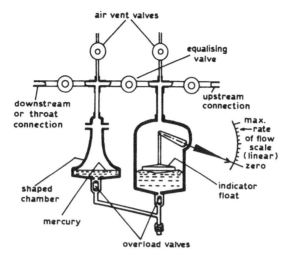

**Fig. 5.7.** Shaped chamber method of square-root extraction.

so that the volume of measuring fluid displaced into the other leg, which contains a float to drive the indicator mechanism, varies. More fluid is displaced and the movement of the indicator is greater when the differential pressure is smaller; the chamber is so shaped that this exactly compensates for the smaller change in differential pressure per unit change in flow rate at low flows.

If a ring balance gauge is used to measure differential pressure across the restricting device, the arm on which the counterweight is hung is shaped so that the moment of the weight about the pivot increases more slowly at first (low differential pressure) than it does later (high differential pressure), thus making the rotation of the ring proportional to the flow rate, rather than the differential pressure (Fig. 5.8).

It is not always practicable to modify the differential pressure-measuring device in order to 'process' the impulse signal. In such cases the differential pressure measurement must be 'linearised' and integrated to obtain total flow.

## 5.6 MEASUREMENT RANGE AND ACCURACY

It is very important with total flow measurement to take into account the effect on the accuracy of the flow *rate*. It is obvious when measuring flow rate that accuracy will suffer if a meter, whose accuracy is a function of

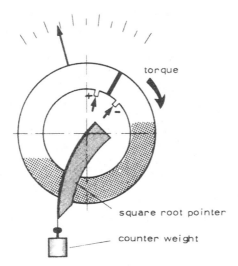

torque

square root pointer

counter weight

**Fig. 5.8.**

*maximum* flow rate (see Section 4.1) is used to measure a flow rate at the bottom end of the meter range. It is not so obvious when measuring total flow that the accuracy still depends on flow rate and not on total flow at all. This is even more difficult to accept since the flow rate may not be measured at all, and may vary with time. For this reason in particular, positive displacement and turbine meters, and other types, the accuracy of which is a function of *actual* rather than maximum flow rate, are best suited to the measurement of total flow.

The most unsuitable methods of flow measurement to use for total flow measurement are those which employ measurement of differential pressure across a restriction device. Nevertheless, since large turbine meters and, even more so, large positive displacement meters are very expensive, it is necessary, particularly when measuring the total flow of gases and vapours, to employ these methods. Not only is accuracy a function of the maximum measurement, but it is a function of the maximum differential pressure measurement not the maximum flow rate. Thus an error of 0·1% of maximum flow rate which, using a positive displacement meter, would become perhaps 0·2% of actual flow rate at 0·1% of maximum flow rate, would become 1% using a linear flow meter (such as a variable area meter), the error for which is a function of maximum flow rate, and would become 10% using a differential pressure system because of the square-law relationship between pressure and flow rate. Nor must it be thought that processing the impulse or transmission signal to 'linearise' the flow/signal relationship, in order to make it possible to integrate it, in any way changes this problem of accuracy; errors in measurement cannot be reduced by signal processing. Obviously, differential pressure measurement across a restricting device cannot be used for total flow measurement unless either the range of flow rates to be encountered is very small, or the accuracy is of little importance, both unlikely conditions to be satisfied.

To overcome this difficulty three different range differential pressure-measuring devices are sometimes used over the 10:1 range in the example given above. Change of measuring device at a little under $\frac{1}{2}$ and $\frac{1}{4}$ of full flow rate, gives an approximately 5:1 range of differential pressure; 0·1% of full scale becomes 0·5% of actual measured flow rate at the lowest flow rate measured by each device, and thus the greatest error over the 10:1 range is 0·5%.

Another way in which differential pressure measurement can be used for total flow measurement in large-diameter pipes is the double venturi system shown in Fig. 5.9. The inlet to the small venturi is located in the low-pressure zone of the large venturi, and so the differential pressure of the 'low flow' transmitter is greater than that of the 'high flow' transmitter. In fact, if the two venturis are dimensionally similar, the low-flow pressure difference will be the square of the high-flow pressure difference. The low-flow transmitter would have to be protected against overpressure, and special provision made to change over the integrator

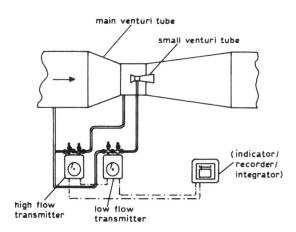

**Fig. 5.9.** Double venturi installation.

to the other range at the high-flow limit of the low-flow transmitter.

The error band and useful range of measurement of both positive displacement and turbine meters depends on the selection of the best size of meter. The wear rate of both types increases rapidly as the flow *rate* increases, but if maximum flow rate will not be sustained a meter can be selected which is smaller and therefore gives accurate measurement over a wider range; if the flow rate will be sustained at near maximum on the other hand, the range is less important. It is usually possible to obtain a range of 10:1 if this is borne in mind.

## 5.7 CALIBRATION OF TOTAL FLOW METERS

When used for accurate measurement, total flow meters often require in-line 'proving' in order to retain calibration accuracy at all times. This can be achieved, as described in Section 4.11, when the total measurement takes a long time and the 'start up' and 'shut down' periods are very short in comparison. However, when the measurement is to be made over a short time period (as for instance in filling a road or rail tanker) these 'transient' periods can cause considerable error. A batch prover tank is then used (Fig. 5.10) to calibrate total meter delivery and take into account these transient errors.

The total volume of the prover tank includes part of the 'swan neck' which, being narrow, provides a well-spaced scale of volume by which to establish any error in total volume. The prover is only suitable, of course, to calibrate a meter for measuring one particular volume.

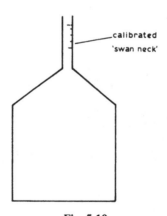

calibrated
'swan neck'

**Fig. 5.10.**

## 5.8 ACCURATE MEASUREMENT OF VOLUME STORED

It is normal to measure total flow and quantity stored in tanks by volume because, as has already been said, it is very much more difficult to measure mass or weight. Nevertheless, since the value of the product is related to weight rather than volume, it is essential to know at what temperature the volumetric measurements have been made. Product stored in tanks may change its temperature by a considerable amount owing to ambient conditions, and for the purposes of inventory control and customs and excise this can create large errors in its value. To overcome this problem, the *average* temperature of the contents of large storage tanks must be accurately measured. There are two ways in which this is commonly done:

1. A 'stack' of thermocouples, each of different length is used to average the temperature over the total depth of fluid in the tank (Fig. 5.11). The averaging is readily achieved by connecting all the thermocouples in series (see Section 6.13). As the level changes in the tank some of them must be disconnected because they are no longer measuring fluid temperature. Thus, the level measurement is used to determine which set of thermocouples are to be connected into the measurement system, and the averaging must take account of the number of thermocouples which are connected at any time.

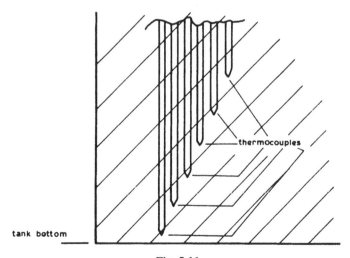

**Fig. 5.11.**

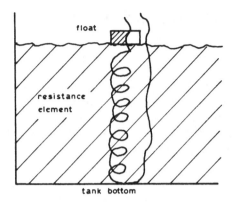

**Fig. 5.12.**

2. A special flexible resistance element is used in some systems (Fig. 5.12). This is attached to a float at the upper end and contracts and expands as necessary to adjust the length of the element to the level in the tank. No switching circuits are required as the resistance element remains unchanged at any level and truly averages the temperature.

## 5.9 TEMPERATURE CORRECTION OF TOTAL LIQUID FLOW

If the product is to be loaded at the point of sale or tax, by measurement of total flow rather than volume loaded into a container (as for instance delivery into the hold of a ship), it is necessary to correct the total *volumetric* flow measurement for variations in the temperature of the product. A continuous weighing system such as that of Fig. 5.13 is often used for this purpose. The product is made to flow through a length of pipe which is suspended from a weighing mechanism and which has flexible connections at inlet and outlet. If the temperature of the fluid changes, the weight of the column resident in the suspended pipe length at any instant will vary accordingly: the signal generated by this mechanism can be used to continuously vary the calibration of the volume flow meter in such a way that it always generates a signal proportional to the volume that would be flowing if the fluid were at some predetermined temperature. If this signal is integrated it will represent total volume corrected to the given temperature.

Correction of liquid flow rate for density variations resulting from temperature changes is not normally possible where accurate measure-

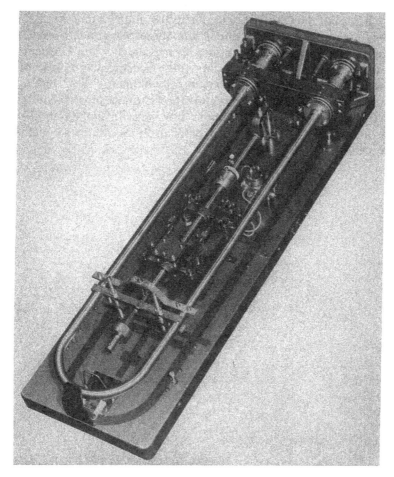

**Fig. 5.13.**

ment is required by measuring the temperature of the flowing liquid, because temperature measurement suffers from delays due to heat transfer whilst flow rate measurement is instantaneous.

## 5.10 PRESSURE CORRECTION FOR TOTAL FLOW OF GASES

Gases suffer change of density not only when temperature varies, but also when pressure varies. Because of this it is not possible to measure total corrected flow to the same degree of accuracy as is possible in the case of liquids; however, since the energy per unit volume of a gas is of a different order to that of a liquid, it is not necessary either. However, the change of density due to change of pressure is often so great that correction is necessary if the volume flow measurement is to be of any practical use at all. Measurements of volumetric flow rate, static pressure and sometimes temperature are fed into a suitable computing device and the corrected volume flow rate integrated to provide reasonably accurate measurement of total volume corrected to 'standard temperature and pressure'.

# CHAPTER 6

# *Measurement of Temperature*

## 6.1 BI-METALLIC THERMOMETERS

Gases, liquids and solids all change in volume (expand) when their temperature increases: one of the simplest temperature-measuring devices consists of two strips of different solid material (usually metals, for strength) joined together along their length. Different materials expand by different amounts for the same change in temperature, and since the two strips of metal are joined together they tend to take up a circular form, with the metal which expands most on the outside in order to accommodate its greater length (Fig. 6.1). The greater the difference of temperature from that at which the bi-metallic strip is straight, the greater the deflection. If one end of the strip is fixed, the 'free' end moves further as the radius of curvature decreases, i.e. the movement of the 'free' end increases as temperature increases.

To provide sufficient movement to operate an indicator mechanism, a long narrow strip of 'bi-metal' is coiled into a helix as shown in Fig. 6.2. In practice several helices are formed, each of smaller diameter than the last, from a single length of bi-metal strip. These are mounted concentrically and arranged in such a way that the longitudinal movement which takes place as the strip 'expands' is in opposite directions in alternate helices, and is thus self-cancelling. In this way an indicating instrument can be built which has a relatively small 'sensitive volume' and low heat capacity. The former is necessary so that the instrument can be inserted into process pipes, the latter so that it will respond to temperature changes relatively quickly.

Such indicators are much more robust than liquid-in-glass thermometers, and can be read more easily. They are widely used for local indication, just as the common Bourdon tube pressure gauge is used for

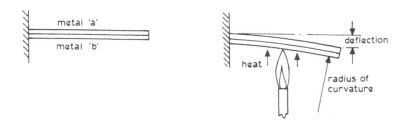

**Fig. 6.1.**

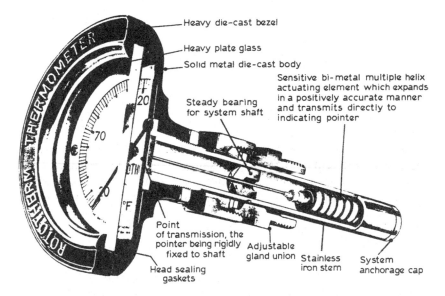

**Fig. 6.2.**

local pressure indication; they are relatively cheap and provide linear indication to an accuracy of $\pm 1\%$ of maximum indication.

## 6.2 LIQUID-IN-METAL THERMOMETERS

Liquids expand very much more than solids as their temperature rises; thus, if a liquid is contained in a metal sheath, a considerable pressure will build up. If this pressure is applied to a pressure-measuring element

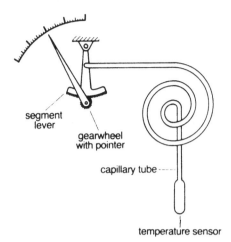

segment
lever

gearwheel
with pointer

capillary tube

temperature sensor

**Fig. 6.3.**

such as a Bourdon tube, the pressure becomes an inferential measure of the temperature. Such a mechanism is shown in Fig. 6.3.

As in the case of the liquid-in-glass thermometer, errors arise unless the whole of the 'fill' liquid is immersed in the process medium; however, this is not possible and normally only the sensor bulb is immersed. Thus, the remainder of the 'fill' liquid as well as the capillary and the pressure-measuring element are 'immersed' in the surrounding air and are sensitive to changes in the ambient temperature rather than the process temperature. Such errors can be compensated for by adding temperature-sensitive elements to the system, which oppose the effects of ambient temperature changes on the main measurement system. A simple example of such compensation is given in Fig. 6.4; a bi-metallic element tends to move the pointer in the direction opposite to that in which it would otherwise move owing to a change of ambient temperature. Since the bi-metal is not immersed in the process material it does not affect normal operation.

This is known as 'case compensation' as the compensating element responds to the ambient temperature of the instrument case. When the range of temperature to be measured is wide, this will usually be adequate, but for narrow-range measurement when the capillary may be subject to a different ambient temperature (close to the plant) from that within the instrument case, this will not be adequate, and a duplicate of the pressure-measuring element and capillary (which is

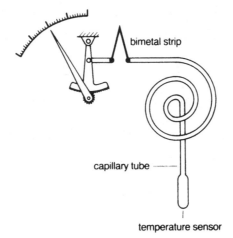

**Fig. 6.4.**

installed alongside the 'real' capillary) are used in opposition to give 'full' compensation (Fig. 6.5). In other words the system is duplicated except for the sensor bulb, and any change of ambient conditions which might affect the accuracy of the measurement of the process temperature is eliminated.

The need for compensation is reduced if the volume of the sensor bulb is large compared to the volume of the capillary and pressure-measuring element. This, however, is not always possible as the bulb must fit into a small-diameter pipe or vessel.

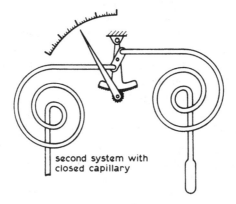

**Fig. 6.5.**

**Table 6.1**

| Liquid | Temperature range in °F | Approximate equivalent in °C |
|---|---|---|
| Mercury | −38 to +1 200 | −39 to +650 |
| Xylene | −40 to +750 | −40 to +400 |
| Alcohol | −50 to +300 | −46 to +150 |
| Ether | +70 to +195 | +20 to +90 |
| Other organic liquids | −125 to +500 | −87 to +260 |

Different liquids are used as 'fill' according to the temperature range and precision required; a stainless-steel capillary is required for mercury fill but cheaper metals are used for other liquids. The whole of the measuring element, bulb, capillary and pressure element must be filled with liquid at the zero temperature and sealed perfectly; therefore the length of capillary must be decided before manufacture. For the same reason the capillary is often protected by a spirally wound steel sheath. Table 6.1 shows the liquids most commonly used.

## 6.3 GAS-FILLED SYSTEMS

The measuring system (bulb, capillary and pressure element) can be filled with gas instead of liquid. Because it is compressible, gas does not generate as great a pressure as liquid for the same temperature change. Response is more rapid for a bulb of the same volume, but the bulb usually has to be much larger in order that its volume is much greater than that of the capillary and pressure element: this is necessary because 'full' compensation is not practicable. The only way errors due to the effect of ambient temperature changes on the capillary and pressure element can be compensated for is by duplicating the entire system, including the sensor bulb (which would remain outside the process). Thus, although gas-filled systems are cheaper to make, they are less accurate and less convenient to use.

## 6.4 VAPOUR-FILLED SYSTEMS (CHANGE-OF-STATE THERMOMETER)

The 'vapour pressure' of a volatile fluid (i.e. a fluid which can co-exist in the form of both vapour and liquid) is dependent upon its temperature.

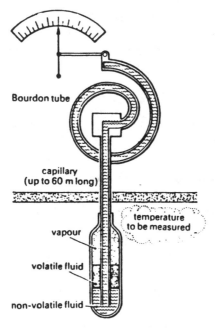

**Fig. 6.6.**

A sensor bulb containing a volatile liquid and its vapour, and also a
more dense and non-volatile transmission fluid (liquid), will develop a
pressure which can be measured by a remote pressure element at the
end of a capillary, as shown in Fig. 6.6.

Expansion of the transmission fluid caused by changes of ambient
temperature has absolutely no effect on the pressure of the vapour,
which depends *only* on temperature: any change of volume is taken up
by a little of the vapour condensing or a little more liquid vaporising.
Thus, there is no need for compensation of any sort, and a very small
bulb can be used (unlike a gas-filled system). The relation of pressure
change to temperature is far from linear; therefore the divisions on the
indicator scale are not of constant size, but become larger as the
temperature increases. Thus, discrimination at the lower end of the
instrument range is poor; as a result vapour-pressure systems are only
used for narrow measuring ranges. An additional disadvantage is that
the sensor bulb must be installed in the upright position, so that the
transmission fluid remains at the bottom. Lastly, since the pressure
developed is much less than it is in a liquid-filled system, the hydrostatic

pressure head of the transmission liquid is not insignificant if the sensor bulb and the pressure-measuring element are located at different levels, and this will introduce calibration errors.

The volatile fluid must be chosen to suit the range of the measurement. This is because, by definition, for a fluid to be volatile it must be well below its critical temperature (at which liquid and vapour are the same). Despite these disadvantages, the vapour system is cheaper to make than a liquid-filled system and avoids the problems of compensation: smaller bulbs than for a gas-filled system can be used, which is often of great importance. Neither vapour- nor gas-filled systems, however, are as accurate as liquid-filled systems or provide as much force to drive indicator or transmission mechanisms.

## 6.5 CORRECT INSTALLATION OF PRIMARY SENSING ELEMENTS

*All* primary temperature-sensing elements have a sensitive area which must be correctly located in the process fluid so that the temperature measured is in fact that which it was intended to measure. The tapping arrangements must, therefore, be carefully designed (see Fig. 6.7).

*Thermowells*
Primary temperature-sensing elements are almost always installed in thermowells (Fig. 6.8) in order that they can be removed for service

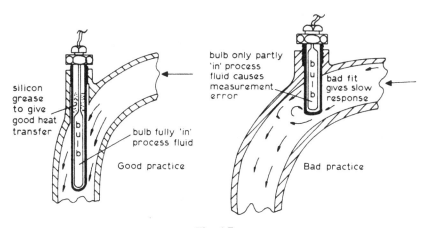

**Fig. 6.7.**

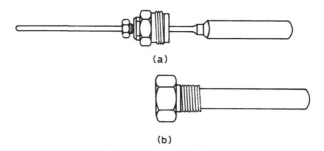

(a)

(b)

**Fig. 6.8.** (a) Temperature-sensing element; (b) Thermowell.

without the necessity of shutting down the process. However, heat has to be transferred from the process through the walls of the thermopocket and through any gas occupying the space between this and the sensing element. Gas is a poor conductor of heat and for this reason thermopockets, though necessary, slow up the response of most temperature-measuring systems, causing considerable difficulty where automatic control is concerned (see Section 8.9). To minimise these difficulties the sensing element is made to be a good fit in the thermowell; for this reason the latter should be manufactured and purchased with the measurement system. Silicone grease or oil should be used, provided the temperature is not too high, to fill the remaining space, as these are much better conductors of heat than gas (air).

## 6.6 THERMOCOUPLE TEMPERATURE MEASUREMENT

If two conductors made of different metals are connected as shown in Fig. 6.9 and one of the two junctions heated to a higher temperature than the other, thermal energy is converted directly into electrical energy, causing an electric current to flow round the circuit. This is

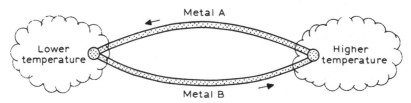

**Fig. 6.9.** The Seebeck effect.

known as the Seebeck effect and can be used as an accurate method of measuring temperature since the voltage between the two junctions (which causes the current to flow) has an exact relationship to the difference of temperature between the junctions.

The two conductors are joined together by welding to form the 'measuring' junction, and usually the other ends of the two wires (of different metals) are connected to form the reference junction. A measuring instrument is connected somewhere in the circuit as shown in Fig. 6.10.

The temperature difference between the measuring and reference junctions is measured. It is, therefore, essential that the temperature of the reference junction should be known, in order to measure the temperature at the measuring junction; this is known as the reference temperature, and it can be established by placing the reference junction in a container of melting ice or at some other known temperature (Fig. 6.11). However, this is generally impracticable and so it is more common to use one of the methods described in the next section to

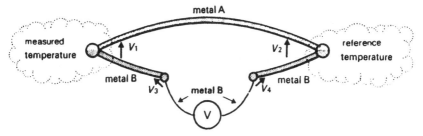

**Fig. 6.10.** A thermocouple circuit showing the contact potentials at the meter connections.

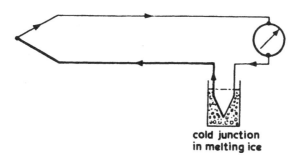

**Fig. 6.11.**

compensate for changes in the temperature of the reference junction, which is usually at ambient.


## 6.7 MEASURING THERMOCOUPLE VOLTAGES

The simplest way to measure the Seebeck effect is to use a galvanometer or current-measuring instrument to measure the current which flows in the circuit. However, the resistance of the leads and even the coil of the measuring instrument will vary with ambient temperature and introduce errors, which then have to be compensated. This is generally done by using a low resistance/high sensitivity galvanometer, and 'loading' the circuit with a 'swamp' resistor, $R_s$, as shown in Fig. 6.12. The swamp resistor can be made from a metal whose resistance changes little with temperature ('temperature stable'); as a result, although the coil of the galvanometer is made from copper, the resistance of which changes considerably with temperature, the total change of resistance is greatly reduced.

*Examples of Cold-Junction Compensation*
In the method illustrated in Fig. 6.13 the reference junction is immersed in a bath of suitable liquid, the temperature of which is controlled at a constant value (above ambient) by a small heater and a thermostat. This too is very clumsy for plant use as the bath must be of a fairly large capacity to provide a constant reference temperature. The method shown in Fig. 6.14 is most commonly used, because it does not require baths of liquid. A 'bridge' is inserted into the circuit at the reference junction, so that it is subjected to exactly the same temperature as the junction. Three of the resistors are made of material (manganin wire) which has a very small change of resistance with temperature, and the fourth of a material (nickel wire) which has a large resistance change

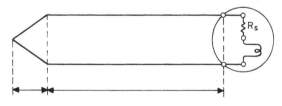

**Fig. 6.12.**

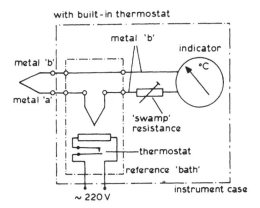

**Fig. 6.13.**

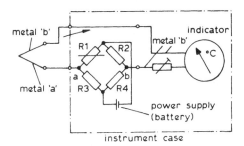

**Fig. 6.14.** Thermocouple circuit showing resistances involved.

with temperature. The resistances are adjusted so that the bridge is in balance at the reference temperature and no voltage difference occurs between a and b. At any other temperature a compensating voltage is generated.

In order to eliminate the errors due to resistance changes in various sections of the system a potentiometer is used for accurate measurements. The principle of the potentiometer is that the voltage generated by the thermocouple is nullified by an accurately known voltage generated inside the measuring instrument: a galvanometer is used only to detect the 'null' point. The principle is shown in Fig. 6.15, where the voltage $E$, generated by the thermocouple, is set off against the voltage drop along the length AC of the slide wire S. The slide wire is itself a resistance, and the position C at which the voltage drop along AC exactly equals $-E$ is established by sliding a moving contact along the line until the

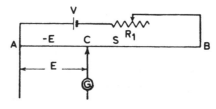

**Fig. 6.15.**

galvanometer indicates no nett voltage. The total voltage drop along the whole length of the slide wire AB is that supplied by the battery or constant voltage source V.

In practice the instrument is 'automated' so that the 'null' point is established by a servo mechanism operating in conjunction with the galvanometer. The circuit of a real instrument would be more complex, and look like that shown in Fig. 6.16. The resistors $R_2$, $R_3$, $R_4$, and $R_5$ perform the reference junction temperature correction ($R_2$, $R_3$ and $R_5$ being made of manganin, and $R_4$ of nickel), whilst at the same time $R_2$ and $R_3$ determine the measuring range of the instrument by fixing the voltages at the ends of the slide wire, A and B. The slide wire shunt resistor determines the span of the instrument. Accuracy of measurement is dependent upon the value of the constant voltage from V. Periodically the galvanometer is connected to the standard reference cell SC by the two-way switch, and the resistor $R_1$ adjusted so that the voltage from V is equal to this reference voltage. In many instruments this is done automatically.

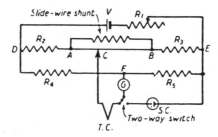

**Fig. 6.16.**  AB, slide wire; $R_1$, adjustable resistance; T.C., thermocouple; G, galvanometer; S.C., standard cell; V, source of supply to bridge.

## 6.8 COMPENSATING LEADS

It is inconvenient and even impracticable to use the two metals of the thermocouple as conductors throughout the system in order that the reference junction shall be at the point of voltage measurement as shown; apart from any other consideration it would mean that much of the internal circuitry of the instrument would have to be constructed of wire of one or other of these materials, and an instrument could only be used with one type of thermocouple. In practice, therefore, connections are made to the instrument as shown in Fig. 6.17.

There are now, however, not two, but three junctions between dissimilar metals, two of them at the reference position. If these two junctions are at different temperatures, an unwanted voltage will be generated which will introduce measurement error; fortunately these two junctions are made within the instrument case and will normally be at the same temperature without any special measures being necessary.

It will be obvious by now that for a number of reasons the reference junction(s) must be located inside the measuring instrument. In theory it (they) could be located in the plant, but in that case reference junction compensation would have to be applied in the plant, and the range of ambient temperatures would probably be greater, making compensation more difficult. In order that the reference junction(s) are at the instrument it is necessary that the conductors (leads) from the

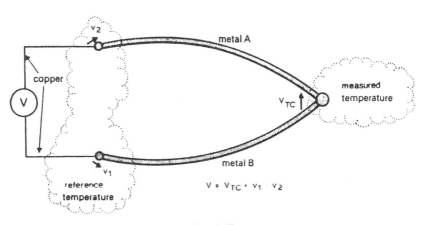

Fig. 6.17.

thermocouple in the plant to the measuring instrument are of the same metals as the thermocouples themselves. Thus, these connections must be made using special cables which are usually referred to as 'compensating leads'. Moreover, when connecting compensating leads, unlike normal cables, it is obviously *essential to connect the correct conductors to each terminal of the thermocouple.*

## 6.9 THERMOCOUPLE TRANSMITTERS

The principles of signal transmission will be covered in Chapter 7 but it is important to note here that the need to use compensating leads from the thermocouple to the measuring instrument has led to the development of a type of instrument known as a 'temperature transmitter'. This is in fact not simply a transmitter, but a measuring and transmitting instrument combined. The very small voltage generated by a thermocouple cannot be used directly to operate a transmission device, but must be amplified first. However, amplification of a direct voltage is not easy, and such instruments are expensive and are often prone to drift problems.

## 6.10 CONSTRUCTION OF THERMOCOUPLES

A thermocouple is simply two wires of different metals joined together to form the measuring junction (usually by fusion or welding). The conductors must be insulated from each other, and in its simplest form a thermocouple is as shown in Fig. 6.18.

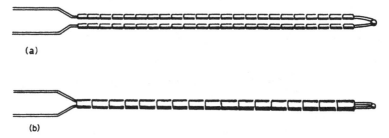

**Fig. 6.18.** (a) Straight-type element; wires with single-hole insulators. (b) Straight-type element; wires with double-hole insulators.

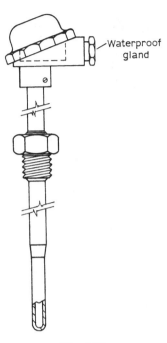

Fig. 6.19.

However, it must be possible to replace the thermocouple itself without shutting down the plant, and it must be protected from mechanical and other damage, for which reasons it is usually installed as shown in Fig. 6.19. The thermowell remains in the process vessel or pipe permanently, and must conform in its construction and installation to all safety and pressure standards which apply to the vessel itself. The sheath in which the 'couple' itself is fitted has its tip pressed, by a spring, onto the bottom of the 'well' so as to ensure good heat transfer. A little silicon oil or grease should be put in the 'well' to further improve heat transfer. The compensating cables are connected to terminals in the 'head', which is waterproof, and enters through a waterproof gland. The 'well' is sometimes constructed integrally with a flange instead of the threaded mounting connection shown.

Using this type of thermocouple, a metal-to-metal contact is essential between the measuring junction and the bottom of the 'well' in order to provide good heat transfer from the process, through the wall of the 'well' and into the thermocouple. This same metal-to-metal contact also

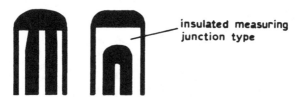

**Fig. 6.20.** Mineral-insulated thermocouple probes.

ensures electrical conductance between the thermocouple and the thermowell, which is itself electrically in contact with the process vessel or pipe, which are in turn normally connected to earth potential. Earth potential can vary considerably throughout the plant: if the measuring instrument is located some distance from the thermocouple, such voltage differences can introduce serious errors into the measurement. To avoid this the instrument can be insulated from earth at the point where it is installed, and a conductor installed to connect it to earth potential at the thermocouple installation point. However, this may interfere with electrical safety (see Section 7.10), and it is preferable to insulate the thermocouple measuring junction from earth potential leaving only the measuring instrument 'earthed'. To do this a special 'sheathed mineral-insulated' thermocouple is used (Fig. 6.20). This thermocouple uses powdered magnesium oxide as an insulator: the measuring junction is insulated from the sheath, so that there is no contact between sheath and thermocouple. Magnesium oxide, however, whilst being a good electrical insulator conducts heat well, so that the measuring performance is good. Insulation between the thermocouple and sheath is at least 1000 M$\Omega$, and so this type of sensor can be used to measure the temperature of the windings of large electrical motors, etc.


## 6.11 MULTIPOINT POTENTIOMETRIC INSTRUMENTS

Self-balancing (servo control) potentiometric instruments are very expensive, and it is therefore common practice to use 'multipoint' forms. A number of thermocouples are connected to the same instrument and are either switched manually to the measuring circuits for indication of whichever temperature is selected by the operator, or are automatically switched in turn for recording purposes. In the recording version the instrument makes a 'dot' on the chart, of a particular colour, each time the measuring system balances in the null voltage state when connected to the appropriate thermocouple. Thus, the recording consists of a series of dots and a number of temperatures can be recorded on a single chart provided they are not too close in value. The number of temperatures which can be recorded by one instrument is also limited by the space between dots, which increases with the number of temperatures, and also by the available colours. In practice it is rarely practicable to record more than 12 points on the same chart.

Multipoint switching can be done by 'single-pole' switching, as shown in Fig. 6.21, in which case only one contact is needed for each thermocouple. One side of each thermocouple is connected by a common wire to the measuring system; this is permissible if earthed

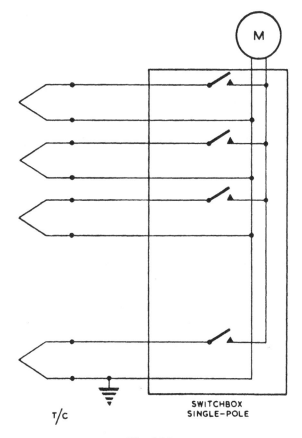

Fig. 6.21.

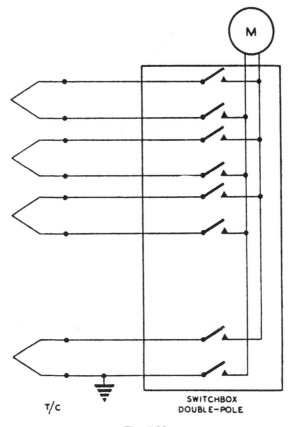

T/C       SWITCHBOX
DOUBLE-POLE

**Fig. 6.22.**

thermocouples are used, only if installed close together on the plant (so that earth potential is not likely to vary significantly). Even so the instrument will have to be insulated from earth potential at the location where it is installed.

In order to avoid these difficulties double-pole switching is used so that the earth potential for the measuring circuit is that of the particular thermocouple being measured (Fig. 6.22). If *all* thermocouples are of the insulated type, single-pole switching can be used and the measuring instrument earthed at its point of installation, which is more convenient. (There are other complications connected with intrinsic electrical safety which will be discussed later in Section 7.10.)

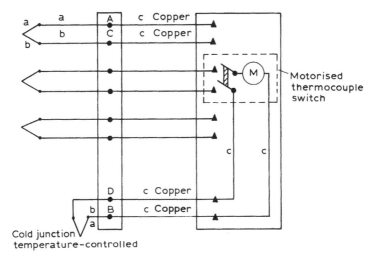

**Fig. 6.23.**

The connections from the terminal board of the instrument to the switches and from these to the measuring circuits are made from copper wire, since an instrument may be used with thermocouples of different materials. The terminal board is therefore the location of the reference junctions, and must be maintained at constant and uniform temperature (see Fig. 6.23 where a and b are the dissimilar metals of the thermocouple).

## 6.12 THERMOCOUPLE MATERIALS

The pairs of metals from which thermocouples are normally made are shown in Table 6.2; graphical plots showing the different relationships between the temperature difference (between measuring and reference junctions) and the voltage generated are given in Fig. 6.24. It can be seen that these types differ greatly in sensitivity, i.e. the voltage produced for each unit of temperature difference. They also differ greatly in their ability to withstand corrosive environments or mechanical stress, and in their precision, i.e. the reproducibility of the voltage/temperature relationship.

**Table 6.2**

| Designated type and British Standard (BS) number | Metal or alloy for 1st wire | Metal or alloy for 2nd wire |
|---|---|---|
| Type S (BS 1826) | platinum | { 90% platinum, 10% rhodium |
| Type R (BS 1826) | platinum | { 87% platinum 13% rhodium |
| Type J (BS 1829) | iron | { constantan (57% copper, 43% nickel plus small amounts of other materials) |
| Type T (BS 1828) | copper | constantan |
| Type E | Chromel (90% nickel, 10% chromium) | constantan |
| Type K (BS 1827) | Chromel | { Alumel (94% nickel, 3% manganese, 2% aluminium, 1% silicon) |

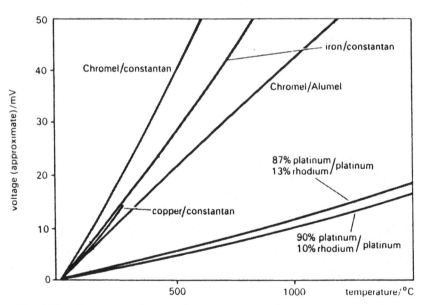

**Fig. 6.24.** Curves of voltage against temperature for several common thermocouple materials.

## 6.13 MEASURING TECHNIQUES USING THERMOCOUPLES

The voltage generated by a thermocouple depends on the *difference* in temperature of the measuring and reference junctions; if both junctions are used for measuring, then thermocouples provide an ideal method of measuring temperature difference as opposed to temperature. It is not unusual to want to measure temperature difference in process control applications; a typical circuit is shown in Fig. 6.25.

In the circuit shown in the figure the two reference junctions are EF and GF, and the conductors AE and DG are of the same material. This is known as 'back to back' connection; the voltages generated oppose each other, and thus represent the difference in temperature of the junctions $T_1$ and $T_2$. The terminal strip EFG can be located anywhere, provided it is at uniform temperature. Changes of temperature of this terminal strip do not matter since both reference junctions rise or fall by the same amount and the change is cancelled out. The conductors EH and GJ can be of any material for the same reason; thus, compensating leads are not necessary.

Average temperatures can be measured very easily using thermo-

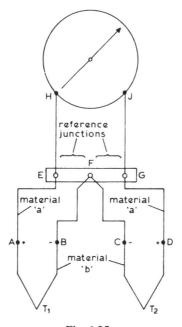

**Fig. 6.25.**

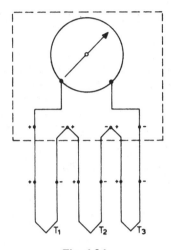

**Fig. 6.26.**

couples; the voltages are simply added together as shown in Fig. 6.26. In this case all the reference junctions must be located within the case of the measuring instrument, in exactly the same way as when a single thermocouple is used. Correction must be applied in the same way, the only difference being that the voltage generated is as many times greater (than it is in a single temperature measurement) as there are thermocouples to average. (The correction voltage is similarly increased.)

Exactly the same arrangement is sometimes used to increase the voltage generated when measuring very small temperature changes — a bundle of thermocouples insulated from each other is used in place of a single 'couple'. This is known as a 'thermopile'.

An important use of thermocouples is to measure the temperature of the surface of pipes which are subject to a corrosive atmosphere and heat, such as the pipes in a pyrolysis furnace. The thermocouples themselves are sometimes laid in a small groove in the surface of the pipe in order to give them protection from direct contact with the furnace radiation (Fig. 6.27). Another method consists of inserting

**Fig. 6.27.**   Thermocouple in shallow groove.

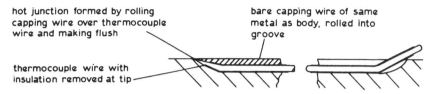

hot junction formed by rolling
capping wire over thermocouple
wire and making flush

bare capping wire of same
metal as body, rolled into
groove

thermocouple wire with
insulation removed at tip

**Fig. 6.28.** Thermocouple arrangement for grooves in copper and steel tubes.

small mineral-insulated 'couples' into holes drilled directly into the pipe wall (Fig. 6.28).

## 6.14 ELECTRICAL INTERFERENCE

The voltage generated by a single thermocouple is very small and in order not to introduce errors the resistances of leads are kept as low as possible. It is very easy for other voltages to be generated in these leads by unwanted 'pick-up' from electrical power equipment on the plant. Such 'pick-up' occurs by transformer effect from power supplies of alternating current, or by actual radio emission of much higher frequency. Fortunately, direct current cannot be 'picked-up' in either of these ways, and so provided precautions are taken to avoid AC pick-up, errors can normally be avoided. These precautions include built-in inductance in the measuring instrument (which will offer no resistance to direct current flow, but oppose alternating current flow) and avoiding the running of thermocouple extension leads near either electrical equipment or alongside electrical cables. Thermocouple leads can also be 'screened' from pick-up by running them in metal conduits or trays which are independently earthed only at one end, and which, therefore, themselves pick up the AC current and conduct it harmlessly to earth without passing through the measuring instrument (Fig. 6.29).

## 6.15 RESISTANCE BULBS

The electrical conductivity of most metals decreases as their temperature increases (see Fig. 6.30). This property has already been seen to provide compensation for the reference junction temperature variations in a thermocouple measuring system, but is also used to measure temperature in its own right. A resistance thermometer bulb (RTB) constructed as

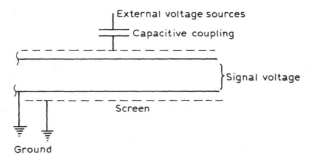

**Fig. 6.29.**

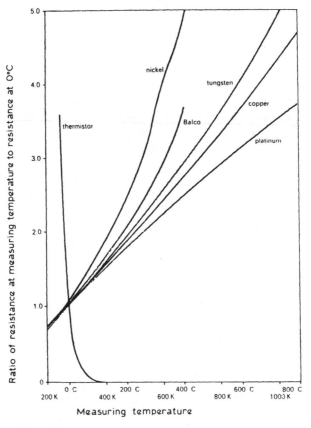

**Fig. 6.30.** (Note reverse slope for thermistor.)

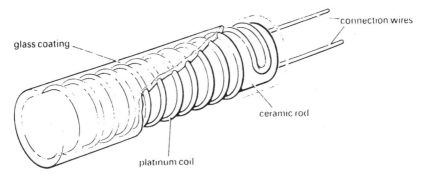

glass coating

connection wires

ceramic rod

platinum coil

**Fig. 6.31.** A resistance thermometer bulb.

shown in Fig. 6.31, is used as one 'leg' in a Wheatstone bridge/galvanometer or cross-coil instrument in one of the ways shown in Fig. 6.32.

In Fig. 6.32(b) the lead to the RTB from one side of the 'bridge' supply is balanced by the lead from the resistance $R_2$ located in the instrument, so that the resistance of the leads is not added to that of the RTB, but evenly divided between the two sides of the 'bridge'. Thus, the bridge is balanced at the minimum temperature of the measuring range, regardless of the length of the connecting leads. If the leads are very long, however, the additional resistance (which also adds to the galvanometer lead) may reduce the sensitivity of the instrument: in such cases it is better to use the system shown in Fig. 6.32(c), even though this increases the cost of the leads still further, there being a total of four conductors. These systems are referred to as two-, three- and four-wire systems.

The measuring systems shown in Fig. 6.32 are known as 'unbalanced bridge' systems, because the resistance bridge is 'unbalanced', producing a voltage across the galvanometer or cross-coil measuring instrument at all except the minimum temperature of the range of measurement. For maximum accuracy a servo-operated 'null balance' instrument is used (as in the case of thermocouple instruments) as shown in Fig. 6.33.

In this instrument the slidewire is driven to the null position by the servo control system, and the galvanometer serves only to detect this 'null point'. This system completely avoids errors due to the calibration of the galvanometer, the heating effect of current flowing through the resistors and any variation in the supply voltage, errors which all occur in unbalanced bridge instruments. This is because the galvanometer is

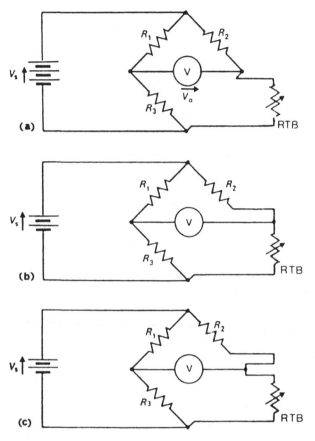

**Fig. 6.32.** Bridges using three-wire and four-wire connections to compensate for resistance changes in the leads.

only used as a detector of the 'null point', there is no current flow at the 'null' state, and the bridge balance depends only on the values of the resistances, not on the supply voltage. Nevertheless, in order to keep the size of the RTB small it is usually made from many turns of fine wire; this is easily heated by a very small current when the system is in the unbalanced state whilst the servo 'finds' the 'null' position on the slidewire. For this reason the galvanometer (or other detector) must be as sensitive as possible so that even very small currents flowing will cause it to deflect. Very accurate instruments use an amplifier as shown in Fig. 6.34 to increase this sensitivity.

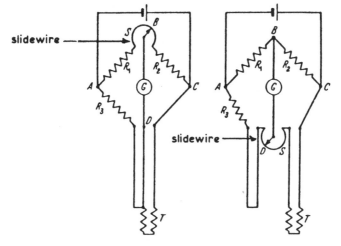

Fig. 6.33.

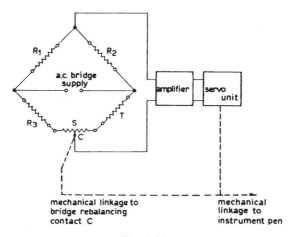

Fig. 6.34.

Cheaper instruments, measuring over a narrower range, often use the very much greater resistance change/temperature change ratio of a thermistor, as shown in Fig. 6.30, to increase sensitivity. Another source of error in resistance bulb measuring systems is variation in the conductivity of the resistances of the bridge circuit (other than the RTB itself). To minimise this source of error these are made of a metal which does not change much in conductivity with temperature, such as manganin; provided ambient temperature changes are not large, this reduces such errors to very small magnitude indeed.

## 6.16 MATERIALS AND CONSTRUCTION

Unlike the thermocouple, the RTB shown in Fig. 6.31 is insulated from earth potential by the glass coating in which the resistance wire is set and there is, therefore, no problem in avoiding errors from earth 'loops'. This glass coating is applied for a different reason, however: the resistance of the wire changes with strain and the construction must be such as to avoid as far as possible any strain in the resistance wire as it is heated.

The RTB is installed in a thermowell in exactly the same way as a thermocouple, and there is exactly the same problem of heat transfer through the wall of the thermowell and sheath into the wire itself.

Compensating leads are not necessary for resistance elements, as it is the resistance which is being measured, not a small generated voltage. Nevertheless, the Seebeck effect will occur at junctions between the metal of the RTB and the leads, which will usually be of a different metal. Care must be taken that the voltages generated are not large enough to cause errors. This will determine that the bridge voltage will be several volts, which in turn determines the order of size of the resistances used, so that the currents in the unbalanced circuit will not cause self-heating problems. The standard RTB usually has a resistance of $100\,\Omega$ at $0°C$ for this reason. The insulation value between the resistance element and its outer metal sheath is of the order of $10\,M\Omega$.

Although it has the smallest ratio of resistance change to temperature increase and is, therefore, the least sensitive, platinum wire gives the most stable relationship and, therefore, provides the most accurate measurement. The other metal most commonly used is nickel, since it gives the largest resistance change. Very high repeatability is possible using platinum resistance elements with null balance instruments — in the order of $±0.1\%$ of span.

## 6.17 MULTIPOINT RESISTANCE-ELEMENT INSTRUMENTS

Multipoint self-balancing instruments are very similar to multipoint self-balancing potentiometric types (except that the bridge circuit

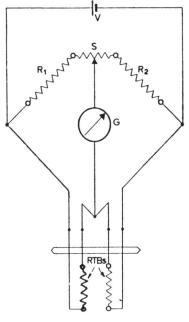

Fig. 6.35.

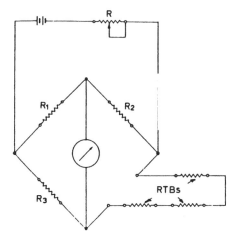

**Fig. 6.36.** Average temperature measurement with resistance thermometers.

differs from the potentiometer circuit). Earthing problems are fewer, as already stated, but care must be taken that the resistance element, which is constructed of coiled wire, does not gain electrical energy by transformer effect from near-by inductive equipment such as electric motors, transformers, etc., as this too will cause error by self-heating.

## 6.18 MEASURING TECHNIQUES

Resistance thermometers can be used to measure temperature difference accurately as shown in Fig. 6.35. Two RTBs are used and are connected into opposite arms of the measuring side of the bridge. Similarly, average temperature can be measured as shown in Fig. 6.36 by connecting several RTBs in series into the measuring arm of a bridge. Alternatively, a special resistance element (in which the coil of resistance wire expands or contracts between a fixed anchorage at one end and a moveable anchorage, such as a float, at the other) is sometimes used for averaging the temperature of liquid in a tank, etc. Care must be taken that the wire is not strained in any way.

# CHAPTER 7

# *Transmission of Measured Data*

## 7.1 INTRODUCTION

The methods of measuring plant or process variables described in the earlier chapters all produce some form of impulse which represents the measured value. This impulse may be a change of pressure in a fluid, a change of position of a float, diaphragm or other mechanism, or an electrical voltage or current. Whatever it is, the impulse is made to operate some form of indicating or recording mechanism located close to the point of measurement. There are considerable difficulties in extending the impulse lines over any significant distance in the case of fluid and electrical impulses, including expense, risk of spillage of flammable liquids and electrical safety. In the case of mechanical impulse there is no question of transmission at all. In the days when control mechanism and measurement indication/recording were located close to the operating location on the plant, there was no problem in most cases, but with increase in size and complexity of plant and the consequent centralisation of both indication/recording and control, it became essential to devise a convenient method to transmit measurement data over much greater distances. This need led to the design of the 'transducer' or 'transmitter', in both pneumatic and electronic forms.

The transducer is a device which generates an 'analog' of the measured value in a form which is convenient for transmission. For instance, if the pressure in a pipe containing air is varied in proportion to the movement of a Bourdon tube, so that the same pressure always corresponds to a given position of the Bourdon tube, then the pressure of the air in the pipe is said to be the analog of the position of the Bourdon tube, and therefore of the pressure measured by the Bourdon

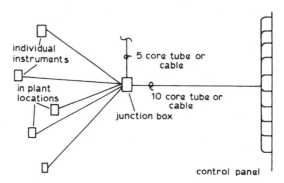

**Fig. 7.1.**

tube. The pipe carrying the air can be extended over large distances in order to transmit this 'analog signal' to some remote location such as a central control room. Compressed air is relatively cheap and thus, provided that the pressures chosen to represent the measured value are not so great as to require expensive high pressure piping or so small that they are difficult to measure at the receiving end, this method of transmission is relatively cheap. In practice a pressure of 1 bar is chosen to correspond to the maximum measured value and multi-core tubes of plastic material are often used to reduce the cost over long distances as shown in Fig. 7.1. Electrical analog transmission is used with electronic instruments employing multicore cables in the same way as multicore tubes for pneumatic transmission. An electric current of 20 mA has now become the almost universal standard to represent maximum measured value, because this is easily measured at the receiving end and is not so large as to require heavy and expensive cables or to make unreasonable difficulties in achieving electrical safety in flammable atmospheres (this subject will be dealt with in Section 7.10).

## 7.2 ANALOG REPRESENTATION OF MEASURED VALUE

For a number of reasons it is not convenient to represent the minimum of the measuring range by either zero pressure or zero current in the analog form; neither is it necessary or even reasonable to do so, since the minimum of a measurement range can be any value, e.g. measurement range 50 to 150 bar pressure. Thus, any value can be chosen to represent measurement 'zero'. In the case of pneumatic analog representation it

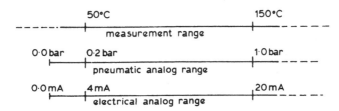

**Fig. 7.2.**

would be very difficult to 'empty' the transmission pipes of air so that the pressure was exactly zero gauge pressure. To try to measure 'zero' in this way is like trying to find out when a tank is empty using level measurement. In the same way it is not practicable to reduce the electric current in an electronic circuit to zero as transistors take current under all conditions. For these reasons a 'live' zero of 20% of the maximum value is used, and the analog signal can therefore exceed maximum and also decrease below minimum measured values during operation, making it possible to be sure that a reading of maximum or minimum on the measuring range is genuine (Fig. 7.2).

It is in fact possible to build an electronic transducer which generates an analog signal with a non-live zero, but only if the power supply is taken to the transducer. This has two disadvantages: first, in addition to the two conductors required to carry the signal current, two additional conductors are required to carry the power supply; secondly, it is almost impossible to make the transducer safe to use in flammable or explosive atmospheres because the electric power cannot be limited to safe levels (i.e. levels which cannot cause a spark to dissipate sufficient energy to cause ignition). Unlike pneumatic transducers, which always have a live zero, many early electronic transducers do not. The 4–20 mA signal range is however almost a universal standard, giving a live zero. Conversion from/to pneumatic signals is thus simplified, since 20 mA corresponds to 1·0 bar (as 4 mA corresponds to 0·2 bar; i.e. $\frac{4}{20} = 0·2$).

## 7.3 PRINCIPLE OF PNEUMATIC TRANSMISSION

When the need arose to transmit measured value signals to remote locations such as central control rooms, early solutions were based on adapting existing local indicating or recording instruments rather than on a totally new design of instrument. Thus, since the existing

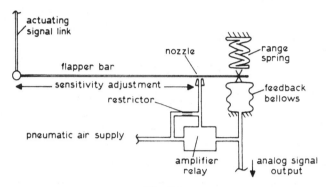

**Fig. 7.3.**

instruments were designed to produce an output in the form of movement of an indicator pointer or recorder pen, the transmitting mechanism was designed to convert mechanical *motion* to a pneumatic analog signal as shown in Fig. 7.3

This type of transducer mechanism is called 'motion balance' for the following reason. Motion of the actuating signal link brings the 'flapper' closer to or further away from the nozzle: relative movement of only 0·02 mm increases the pressure in the pipe behind the nozzle by the full range of 0·8 bar, because it obstructs the air flow through the nozzle (the nozzle is only a small hole and air flows through it at high speed). This increased pressure expands the feedback bellows, moving the flapper in the opposite direction until the higher pressure in the bellows is in balance with the spring force. The actual increase in pressure which corresponds to a given *movement* of the actuating link depends on the spring 'rate' and the position of the nozzle between the two pivots (one at the actuating link connection and the other at the range spring). This type of mechanism is called a negative feedback servo mechanism, and it provides very accurate and repeatable correspondence between the input (the position of the actuating link) and the output (the analog signal).

## 7.4 AMPLIFIER RELAYS

The function of the amplifier relay is to greatly increase the *flow* of air (not the pressure), so that the speed at which the feedback servo action responds is very fast. The flow rate of air through the nozzle is very small and the time it would take for a change in this flow rate to 'fill' the

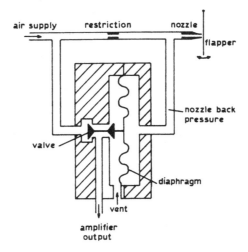

**Fig. 7.4.** Continuous bleed relay, direct acting.

capacity of the bellows and the transmission pipelines would be far too long in practice. The relay (shown in Fig. 7.4) operates as described below.

In equilibrium (or steady state) the pressure behind the nozzle will be such that the flow of air through the restriction is exactly equal to that through the nozzle. If the flapper moves towards the nozzle, reducing the flow, this pressure will increase, as the air flowing out of the nozzle will then be less than that flowing in through the restriction. The increase of pressure will move the thin diaphragm to the left and restrict the flow of air which is passing through the left-hand chamber to the exhaust port through the conical valve. Air flows freely to the output to 'fill' the feedback bellows and transmission pipes until the pressure in these has risen sufficiently to partially restore the thin diaphragm to its original position — negative feedback servo action. Hence, the relay itself operates by negative feedback servo action and forms part of the 'motion-balance transmitter mechanism' which also operates on the negative feedback servo principle. The diagram in Fig. 7.5 shows how the transmitter mechanism is added to an existing indicating mechanism.

## 7.5 FORCE-BALANCE TRANSDUCERS (PNEUMATIC)

In more recent times transducers have been designed which are not indicating or recording instruments; these are known as 'blind'

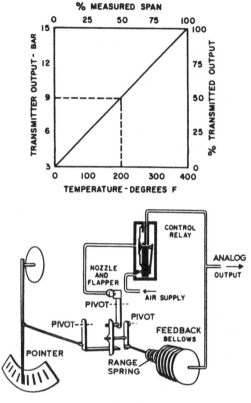

**Fig. 7.5.**

transmitters. Because there is no need to operate a pointer or pen over an indicator scale or recording chart, these blind transducers can be designed on a different principle known as 'force balance' (local indication can still be provided by pressure indicators or milliammeters which measure the analog signal value). Negative feedback servo action is still used to generate an analog representation of the measured value with precision and speed of response, but the impulse *force* is used, not a movement generated by this force through a mechanical mechanism, which often introduces error in the form of lost motion, friction, etc.

It can be seen by comparing Fig. 7.6 with Fig. 7.3 that a very small movement of the beam at the measurement impulse end will be required to cause an increase in pressure at the analog signal output. It can also be seen that the impulse force is opposed by the force generated

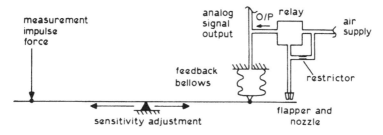

**Fig. 7.6.**

in the bellows by the analog pressure — hence, the name force balance. It should also be noted that the feedback movement of the bellows does not serve to increase the movement required at the measurement impulse end as it does in the case of the motion-balance mechanism.

Probably the most commonly used force-balance transducer is the differential pressure (DP) transducer shown diagrammatically in Fig. 7.7. The force bar transfers the force developed by the capsule onto the lever arm which carries the feedback bellows via the 'flexure strap'. The sensitivity adjustment shown in Fig. 7.6 takes the form of the range wheel, which allows the span of the transducer to be altered to suit a wide variety of applications. The 'zero adjustment spring' holds this fulcrum against the frame of the instrument as well as allowing zero to be adjusted.

The instrument shown in Fig. 7.8 is a force-balance blind transducer using a liquid-filled measurement system to transmit temperature. Note that the relative positions of input force, feedback bellows and flapper/ nozzle are the same as in Fig. 7.6: compare this with the relative positions on the DP cell above.

An interesting example of a force-balance transducer is to be found in the target flow-rate meter shown in Fig. 7.9. This can be regarded as an orifice plate meter in which the 'hole' and the 'plate' have been interchanged. The target causes a restriction and the difference in upstream and downstream pressure acts across this to provide a force which is proportional to the square of the flow rate, just as the differential pressure across an orifice plate is proportional to the square of the flow rate. The force acts through a force bar in the same way as the force developed by the capsule in the DP cell; the rest of the mechanism is also virtually the same. The target flow meter is interesting as an example of a measurement device which would not be possible without the principles of force-balance negative feedback servo action.

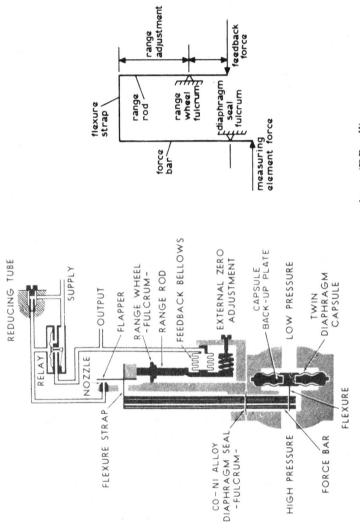

**Fig. 7.7.** Differential pressure transducer (DP cell).

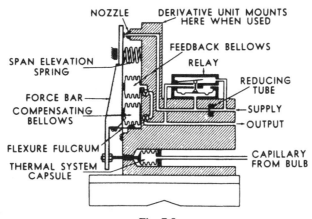

**Fig. 7.8.**

There are many examples of force-balance pneumatic measurement systems, but they all have in common the fact that indication or recording can only be achieved by measurement of the analog pressure, i.e. the transducers are blind. This is no disadvantage when remote central control rooms are used to locate all indicators, recorders and

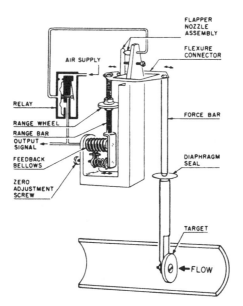

**Fig. 7.9.** Target flow transmitter.

controllers, and local indicators are often added as additional 'readouts'. In fact, one advantage of blind transmitters is that any reasonable number of indicators or recorders can be added to any transmitter.

## 7.6 FORCE-BALANCE TRANSDUCERS (ELECTRONIC)

Electronic transducers operate in exactly the same way as pneumatic transducers; however, the analog signal generating mechanism consists of a negative feedback electrical servo system as shown in Fig. 7.10. A power supply located in a remote position (usually an electrically safe area) provides a regulated voltage at one end of the two-wire current analog circuit, and supplies current in the range 4–20 mA according to the impedance of the transducer. This impedance is made up of two constant elements, the inductance of the 'force coil' and that of the oscillator, together with a variable impedance, the DC current regulator unit. The AC current generated by the oscillator (from power drawn from the analog DC current circuit — hence, the need for the 'live' zero) is fed to a bridge circuit, two elements of which are the differential plate capacitance shown in the diagram. Increase in measurement impulse force causes the centre plate to move towards the upper plate and away

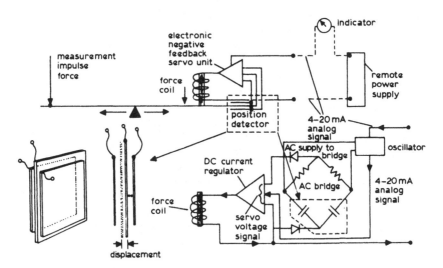

**Fig. 7.10.**

from the lower one, unbalancing the bridge and raising the voltage fed to the DC current regulator. The impedance of the DC current regulator is reduced, increasing the current in the analog circuit and hence the force exerted by the force coil, which opposed the measurement impulse force. The position of the centre plate of the differential capacitance is thus restored almost to its original position by negative feedback servo action in exactly the same way as the flapper and nozzle are in the pneumatic transducer. The voltage generated by the bridge and therefore the change in analog current and the force exerted by the force coil for a small movement of the centre plate is much greater than that required to restore it, just as the pressure generated in the feedback bellows of the pneumatic version is much greater than is needed to restore the flapper to its original position. The feedback amplification factor (or gain) is thus very great, ensuring that the final movement of the flapper or the capacitance centre plate is extremely small.

Other devices are used by some manufacturers as position detectors, e.g. the phase-sensitive detector shown in diagrammatic form in Fig. 7.11. Like the differential capacitor, it requires an AC power supply, but will produce quite a high voltage output for a very small movement.

## 7.7 NON-FORCE-BALANCE TRANSDUCERS

The advantage of the force-balance principle is that the characteristics of the amplifier are not important; the current developed in the force coil or the pressure in the feedback bellows produces a force which is such as to balance the force applied by the measurement sensor at all times, by servo feedback. The relationship between the measured value and the analog signal current or pressure is not subject to drift and is

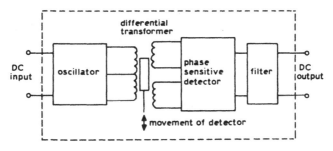

**Fig. 7.11.**

repeatable within ±0·1% of span. However, it is not always possible to apply the force-balance principle to the generation of analog signals, as the impulse produced by some measurement sensors is not a force. Obvious examples are thermocouple and RTB temperature sensors: it will be recalled from Chapter 6 that high measurement accuracy is obtained by the use of a potentiometer, and generation of an analog signal from such an instrument would necessitate a motion-balance mechanism. The lost motion in pivots, etc., would degrade the accuracy obtained by using a potentiometer. The variable voltage impulse generated by a thermocouple, or (together with a constant voltage supply) a resistance element, cannot be transmitted over normal cables without considerable loss of measurement accuracy, and so a transmitter is required. The voltage is amplified and can then be used to generate a 4–20 mA analog without serious error; however, unlike a force-balance mechanism, the relationship between measured value and analog signal current is very dependent upon the gain of the amplifier. This must not therefore change with ambient temperature or simply the age of the components.

To avoid this a 'differential' amplifier is used, the circuit of which is shown in Fig. 7.12. As the voltage on the base of the left-hand transistor increases, the current through the emitter/collector from the positive to the negative rail increases. The output is the voltage between this collector and the collector of the right-hand transistor; because the current through the left-hand transistor, and therefore its collector resistance, has increased whilst that through the right-hand transistor and resistance has not, the output voltage will increase. However, any change in either the transistors or resistors owing to ambient

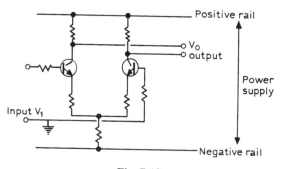

**Fig. 7.12.**

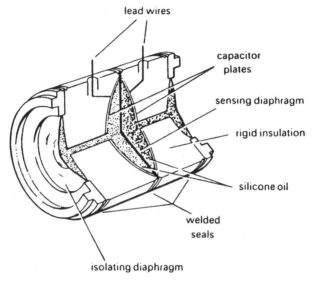

lead wires

capacitor
plates

sensing diaphragm

rigid insulation

silicone oil

welded
seals

isolating diaphragm

**Fig. 7.13.**

temperature variation or ageing will be the same provided: (1) they are matched to start with; (2) they are mounted on the same heat sink; and (3) they are of identical construction. Thus, such amplifiers are not significantly subject to drift. Several such stages of amplification may be used in one transmitter, and today such amplifiers are available in silicon chip form.

In addition to thermocouple and RTB transmitters, high-stability amplification of an electrical impulse is used to generate analog signals from other sensors which do not produce a force or pressure impulse. For example, the differential pressure sensor shown in Fig. 7.13 operates on the principle of the differential capacitor, which is used as the position detector in the force-balance transmitter. Increase in pressure at one end of the capsule with respect to the other causes the sensing diaphragm to move nearer to one of the fixed plates and thus further away from the other. This produces a change in voltage in the bridge circuit in exactly the same way as in the position detector; this is then amplified to generate a change in analog current. However, there is no feedback system and the amplifier must have a stable gain if the errors between the analog and the true measured value are not to be large.

## 7.8 DIGITAL TRANSMITTERS

The analog transmitters described so far all require an impulse input from the measurement sensor, either in the form of a mechanical force or an electrical voltage. Some important methods of measurement, such as the wire and pulley level measurement system, produce neither: nor would it be possible to translate the long movement of the wire or the many revolutions of the drum into a force or voltage with any accuracy. For such applications a digital representation of the measurement is both more appropriate (easily displayed) and much more accurate. The shaft of the drum is fitted with an 'encoder disc' (Fig. 7.14).

A number of light sources shine through the clear portions of the disc or are blocked by the dark portions, as shown diagrammatically in Fig. 7.15. As the disc rotates by an amount equal to the distance between the outermost markings, one or more of these light sources is either blocked or uncovered. In a corresponding set of photosensors each photosensor produces a voltage or does not according to whether the light source with which they are associated is blocked or not. Thus, for any position of the disc there is a unique set of outputs which is a coded description of the position. Measurement discrimination depends on the number of markings in the outermost band and the number of revolutions the drum makes over the measurement range. The number of markings in the outer band depends on how many bands (how many light sources) there are in the transmitter. Any code can be used,

**Fig. 7.14.**

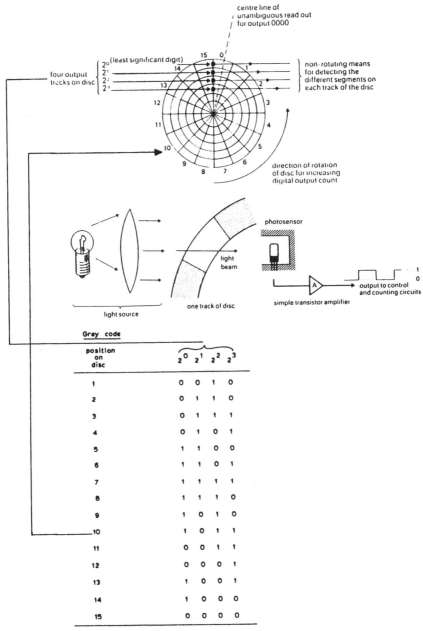

**Fig. 7.15.** The principle of the digitally encoded disc (four output tracks per disc).

provided each position to be discriminated has a unique combination of light sources obscured or not, but in practice the code which is most common is the 'Grey' code shown for four lamps (Fig. 7.15; the disc in Fig. 7.14 is for 10 lights) which is devised so that as the disc turns only one light changes at a time. The coded outputs are decoded by electronic decoding 'logic' and drive a digital display of the measured value.

## 7.9 TELEMETRY

So far in this chapter measurement transmission over distances up to about half a kilometer has been considered, i.e. the distance from a plant-mounted transmitter to the central control room. However, in some cases, e.g. pipelines, measurement data have to be transmitted over much longer distances — sometimes many miles. The cost of pneumatic tubing or single electrical cables would be unacceptable, as would the pressure loss or voltage drop over such distances. Other means have to be found which are cheaper, and these are generally referred to as telemetry.

If, as is often the case in process industries, it is not necessary to transmit data quickly, a simple method called pulse duration modulation is used; a tone on a telephone line, for instance, can be switched on and off during a given period of time which is made to represent the full scale value of the measurement. Thus, if the tone is switched on for the first half of this time interval and off during the second half, the measured value is one-half full scale (Fig. 7.16(a)).

Another similar method is called pulse position modulation. Again, an interval is chosen to represent full scale, but in this system a pulse of short duration is generated after an interval of time which is the same fraction of the full interval as the measured value is of full scale (Fig. 7.16(b)).

These methods require only simple converters to generate the pulse signals, but are only suitable for relatively slow data transmission over distances of a few kilometres. In the same way that processes store energy and thus cause delays (for instance, heat exchange), so also do transmission lines store electrical energy in the capacitance and inductance they represent. The signal pulse is in fact a change of energy; some of this energy is stored temporarily in the transmission, rather like a bucket of water poured quickly into a bath is stored temporarily and only flows away down the waste pipe gradually. Just as the water reaches

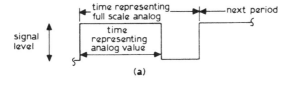

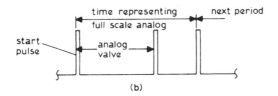

**Fig. 7.16.**

the outside drain more slowly than it is poured into the bath in the first place, so the energy of the pulse reaches the receiving end of the transmission line more slowly than it is generated at the transmitting end (Fig. 7.17). It is difficult to be sure exactly when the pulse begins, and as a result errors may occur.

If the analog signal is converted into a digital code this rounding off of the pulse leading and trailing edges matters very much less. In fact, provided the receiving equipment can reliably detect the presence or absence of a pulse, there will be no loss of accuracy at all! Data values can be represented as a group of bits (binary digits) each of which has a value 1 or 0 to represent the binary equivalent of the value to be transmitted. Each bit is then transmitted over the 'data link' or transmission line during a fixed interval of time as shown in Fig. 7.18.

To do this the receiving equipment must be synchronised with the transmitting equipment so that it recognises the intervals of time correctly. This is achieved in one of two ways:

1.  Each 'byte' or group of bits will comprise a fixed number of time intervals. At the beginning of each byte the transmitter will always send a '1' or 'mark' signal. The receiving equipment will be

**Fig. 7.17.**

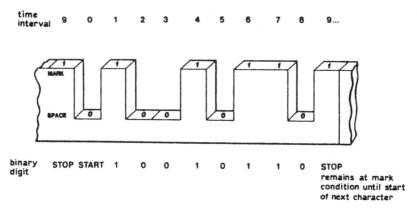

Fig. 7.18.

designed to start its timing of the intervals making up the next 'byte' as soon as it detects the end of this mark (which is not part of the data). Its own internal timer will be accurate enough to determine the end of each interval throughout the byte, at the end of which it will receive a further synchronising 'mark'. This is known as 'asynchronous' transmission (Fig. 7.19(a)).

2. Clock pulses are transmitted together with the data bytes over a separate channel, so that the receiving equipment always operates in synchronism with the transmitter. This is known as synchronous transmission (Fig. 7.19(b)).

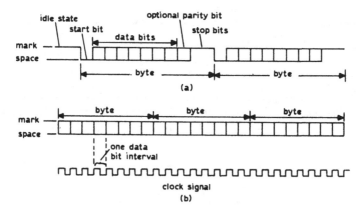

Fig. 7.19. (a) Asynchronous and (b) synchronous transmissions.

## 7.10 INTRINSIC SAFETY

This is a subject in its own right, but it is essential that the basic requirements as they apply to 4–20 mA analog transmitters are understood. The principle of intrinsic safety is only possible in conjunction with large centralised control rooms which can be made 'safe areas'. On a refinery or other plant, where there is a risk that flammable vapours may escape, it is normal to designate areas of the plant safe, possibly dangerous or dangerous, according to the risk of such vapours existing in dangerous concentrations. A control room, regardless of where on the plant it is located, can be made safe by pressurising it very slightly; in this way it is not possible for gases or vapours to enter from outside. The supply of pressuring air *must* of course be taken from a safe area (usually at some height above the ground) and other precautions, such as high-level cable and services entry (to avoid the possibility of vapour entering through cable ducts) and effective air locks at doorways, are essential.

Intrinsic safety means that the measurement and transmission (and control) system must be designed, constructed and *maintained* in such a way that it is not possible for sufficient energy to exist in that part of the system which is outside the safe area to cause ignition or explosion (on the assumption that flammable or explosive mixtures *do* exist). This in turn means that the transducers and cables must not have enough electrical inductance or capacitance to store the critical amount of energy, nor must it be possible to transfer this amount of energy into the 'hazardous area' from the 'safe area' in an interval of time equal to that in which a spark endures. The amount of energy required to ignite a mixture of any particular gas with air is extremely small, but varies with the gas or vapour. The transducers themselves must be designed to satisfy the requirements laid down by suitable national certification authorities, who have the results of considerable research into the energy of ignition to draw on. In addition 'barriers' are installed in cables leaving the 'safe area' to ensure that the energy transfer rate is limited to a safe value. These barriers which are normally of the passive type (that is, they consist only of resistors, fuses and diodes) are intended to provide a low resistance path to earth if a fault occurs in the power supply side, i.e. before the cable leaves the safe area. Figure 7.20 illustrates the principle of operation. If the current (flow of energy out of the safe area) rises, the resistances cause the voltage to rise at the diodes which conduct the energy harmlessly to earth inside the safe area. The

140

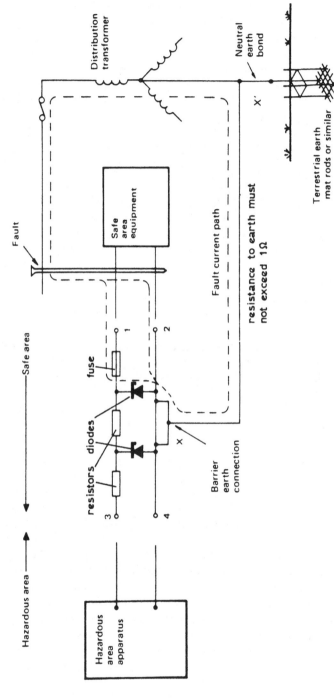

**Fig. 7.20.**

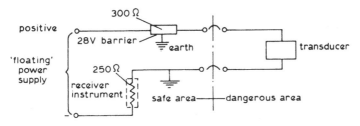

**Fig. 7.21.**

second diode and resistor are there in case the first fail, and the fuse will normally 'blow' to prevent overheating should a fault develop.

A diagram of the same circuit showing that one of the conductors in the cable to the transducer has to be connected to earth is given in Fig. 7.21. Because the receiver is connected between the earth connection at the barrier and the power supply (in order that it is located in the safe area) the power supply has to be 'floating'. This means that as the current flow in the circuit (the analog signal) varies, the positive terminal of the power supply increases in voltage, whilst the negative decreases with respect to earth. This has the disadvantage that a common power supply cannot be used for a number of 'loops' which would be less expensive than a separate power supply for each.

The solution is to use two barriers as shown in Fig. 7.22. In fact the lower barrier can differ from the upper; it can have lower resistance values. Two barriers (one 28 V and one 10 V) are commonly constructed in one unit for this purpose. The disadvantage of this solution is that it limits the impedance of the transducer, and some manufacturers have difficulty keeping within this constraint when designing the equipment. The reason for this is easy to see: at a maximum analog current flow of

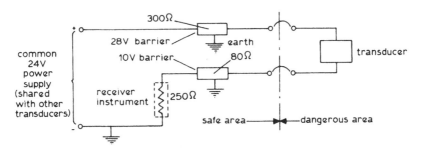

**Fig. 7.22.**

20 mA and a supply voltage of 24 V (the 28 V barrier diodes start to conduct at about 26·5 V) the total loop resistance must not exceed

$$R = \frac{V}{I} = \frac{24}{0 \cdot 02} = 1200 \, \Omega$$

It can be seen from Fig. 7.22 that, excluding the transducer, the loop resistance is 250 + 80 + 300 = 630 $\Omega$, which leaves only 570 $\Omega$ for the transducer. Since the function of the transducer is to modulate (regulate) the current flow, it can be regarded simply as a variable impedance in the circuit. Thus, when a 20 mA analog current flows in the circuit, the transducer adjusts to offer an impedance of 570 $\Omega$, whilst at 4 mA current

$$\text{total } R = \frac{24}{0 \cdot 004} = 6000 \, \Omega$$

Hence

$$\text{transducer impedance} = 6000 - 630 = 5370 \, \Omega$$

# PART 2

# CHAPTER 8

# *Principles of Control*

## 8.1 INTRODUCTION

Previous chapters have been concerned entirely with techniques of measurement of process variables and with the transmission of signals representing the measured values. Such measurement systems are essential in order that the process and the plant operations can be controlled, whether such control is carried out by the human operator or by automatic control systems. In most cases control is implemented by manipulating the position of a throttling valve in a pipeline, so as to increase or decrease pressure, flow rate or temperature. The valve is called the 'final control element'; other final control elements can be speed control systems on pumps or prime movers, or any other device which manipulates energy or material flow. Early in the history of industrial development it was realised that it was an inefficient use of human skills for a human operator to have to set the position of the final control element to a new value every time a change in the process conditions occurred. As a result mechanisms were developed which could regulate the flow rate, pressure or temperature at any value the human operator set (the set or desired value) even when process conditions changed (disturbances). These mechanisms used the same feedback servo control principles that were described in Sections 7.3–7.6 to match the measured value to the desired value by either opening or closing the control valve as appropriate. However, this is a much more difficult control problem because the process variable cannot usually be made to change quickly, unlike the transmission signal; care has to be taken to 'tune' the control action to suit the way the process can be made to 'respond'. Initially, these mechanisms were constructed locally — they developed before signal transmission — as an extension of non-

transmitting (impulse indication) instruments. Later, as signal transmission was developed and process plants grew too large for the
operators to control them from local scattered control rooms, the
control mechanisms were transferred to a centralised room remote from
the final control elements so that the operator no longer had to go out on
the plant to adjust a flow rate, temperature or pressure setting.

The term 'process control' covers *all* aspects of the control of each
process and the operation of the plant within which the processes take
place. In a modern process plant, such as a refinery, this always includes
many of the control mechanisms described above. It is very important to
realise that the human operator still controls the process, with the help
of these mechanisms, though it is also true to say that it is no longer
possible to operate any modern process plant without many such
mechanisms. These mechanisms *do not automate the plant*; development
of automated plant is just beginning. The design, construction and
operation of such mechanisms is control technology.

## 8.2 NATURAL REGULATION

The system shown in Fig. 8.1(a) will respond to a change in the inflow
rate in a stable manner; the level will change until the hydrostatic head *h*
is again just sufficient to overcome the resistance of the valve at an
outflow rate exactly equal to the new inflow rate. In other words, the
system will come to another state of equilibrium without the need for
any control action at all. Such a system is said to have natural
regulation. Only if it is necessary to maintain a constant level is any
form of control necessary (Fig. 8.1(b)).

The system shown in Fig. 8.2, however, does not exhibit natural
regulation. If there is any change in the inflow to the system (Fig. 8.2(a))
the tank will either empty or overflow eventually; the system is unstable

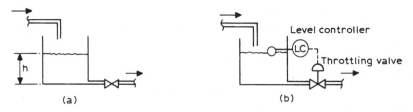

**Fig. 8.1.**

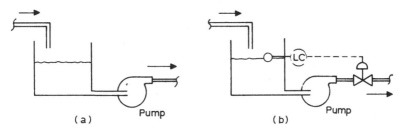

**Fig. 8.2.**

without control. It is obviously necessary to add a control loop as shown in Fig. 8.2(b) in order to operate the system at all. Obviously the control action is more difficult to define in the second case.

## 8.3 ON–OFF CONTROL ACTION

Control action must oppose the disturbance: the simplest control action is 'on–off'. In this form either the final control element is in one extreme position or the other (fully open or fully closed), according to whether the error (the difference between the measured and desired values) has a positive or negative value (Fig. 8.3). However, this form of control action is always too great because it is not proportional to the size of the error,

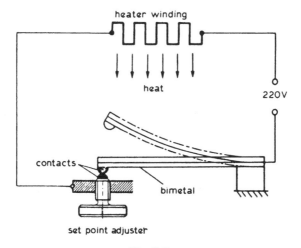

**Fig. 8.3.**

but always a maximum. Even more important, delays in the system will nearly always ensure that changes of control action are 'out of phase' with changes in the controlled variable. Control action will only be applied at all after an error has developed. This combination of excessive control action and poor timing makes on-off control action suitable only for the simplest and least critical applications, such as space heating, where the process exhibits very considerable self-regulation and large capacity.

## 8.4 PROPORTIONAL CONTROL ACTION

Nearly all industrial control is of the 'modulating' type, in which a final control element 'manipulates' the process variable (usually flow rate of a fluid) to proportion the control action to the size of the error. Control action is thus never excessive in magnitude although it may still be out of phase with the process disturbance. Two examples illustrate this principle.

*Example 1: Gas Pressure Regulator*
With the gas pressure regulator (Fig. 8.4) a unique value of the correcting variable ($y$) corresponds to every value of the controlled variable ($x$). The movement of the graduated indicator is equal on both sides when the pivot of fulcrum (L) of the lever is exactly in the middle of the lever (Fig. 8.5). If the pivot is moved to the right, small movements of the actuating piston result in larger movements of the final control element; hence, proportionally greater control action results (Fig. 8.6). This effect is called 'amplification'.

The amplification factor ($V$) is determined by the choice of the position of the fulcrum (L). As can be seen in Fig. 8.6 the factor may be greater or less than unity.

$$V = \frac{y}{x}$$

Displacing the fulcrum to the left means that large $x$ values will yield small $y$ values. Strictly speaking, it is not a matter of amplification here, but rather of attenuation (Fig. 8.7).

*Example 2: Level Control*
If the lever arms $a$ and $b$ in Fig. 8.8 are of equal length, the final control element will move 1 cm when the water level moves 1 cm. Thus, the

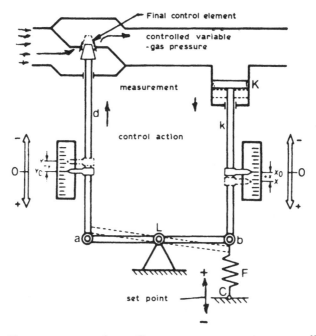

**Fig. 8.4.** Gas pressure regulator. Greater gas pressure causes smaller valve opening and therefore reduces flow of gas. Spring force (F) opposes the gas pressure. 'Set' pressure is increased or decreased by moving 'C' up or down to adjust spring force.

maximum inflow of water will be obtained when the final control element is wide open, whilst flow will cease altogether when it has moved a distance equal to the inside diameter of the pipe. Hence, the control range is equal to the diameter of the pipe; in other words, control action varies from maximum to minimum when the level in the tank varies by an amount equal to the diameter of the pipe. If, however, the fulcrum point is moved to left or right so that the ratio of *a* and *b* is less than or more than unity, the range of level change corresponding to full control action will be greater or less, respectively. The ratio *a/b* is called the *proportional action factor* of the control mechanism.

## 8.5 INTEGRAL CONTROL ACTION

Suppose that in Example 2 in the previous section the valve *z* is opened, increasing the outflow from the tank. If the level is to be maintained, the

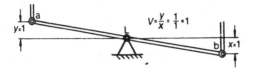

**Fig. 8.5.**

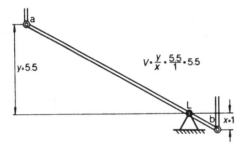

**Fig. 8.6.**

**Fig. 8.7.**

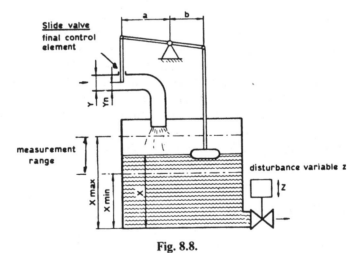

**Fig. 8.8.**

slide valve at the inlet (the final control element) must be opened to increase the inflow until it is equal to the outflow rate. However, in order to achieve this the float must fall sufficiently to open the slide valve enough to increase the inflow by the required amount. Therefore, when the system is restored to equilibrium, the level must of necessity be lower than before the disturbance; in other words complete regulation of the level is not possible using proportional control action. There will always be a change of controlled variable for every change in the disturbance variable, though it will be less than if there were no control mechanism (only natural regulation). The greater the proportional action factor, the less such 'offset' will be. However, if there are delays in the control loop, it may not be possible to use a high proportional action factor without causing instability and so in many systems it is necessary to define some other form of control action if offset is to be avoided.

In the system shown in Fig. 8.9 it is the voltage applied to the motor (M) which varies as the level changes in the tank; the position of the slide valve only *begins* to change when the level changes and the *rate* at which it changes is proportional to the change in voltage and therefore to the change in level. This is called 'floating' or 'integral' control action: at any instant the control action is the sum of the movement of the slide valve since the time when the measured value was last equal to the set point. (In mathematical terms it is the 'time integral of the error', in exactly the same way that total flow is the time integral of the flow rate. Hence, the name 'integral action'.) Since control action continues to change whenever there is an error, it follows that, once the disturbance

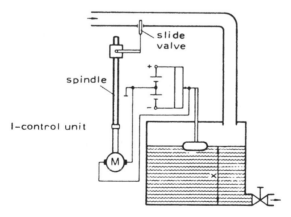

**Fig. 8.9.**

variable stops changing, the level will (eventually) be restored to the desired value without any offset. However, since the position of the slide valve only starts to change when the disturbance variable changes (while proportional control action is immediate), a further delay has been added to those of the process and measurement, and the danger of instability is increased.


## 8.6 PROPORTIONAL PLUS INTEGRAL CONTROL ACTION

It hardly needs stating that better control can be obtained by a combination of the two previous control actions. Proportional action provides an immediate correction which, provided the proportional action factor is not too large, is less likely to make the system unstable, while integral control will eventually remove the offset which cannot be avoided with proportional action alone. For this reason integral control action is often referred to as automatic reset. The two previous mechanisms are combined in the diagram given in Fig. 8.10. Control action is added at the 'summing point', but notice that the fulcrum is now at the top of the spindle: the proportional action factor is therefore the ratio of the lever arm from the summing point to the fulcrum over the total length of the arm $a/b$, and will always be less than unity for this mechanism.

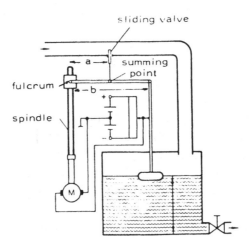

**Fig. 8.10.**

The rate at which the motor is driven, and therefore the rate of integration, can be varied and this is defined as the integral action factor. For convenience the integral action factor is numerically defined as the time taken for the integral control action to move the final control element the same distance that proportional action would move it for a given error, and therefore it changes with the latter. It should be appreciated that under proportional plus integral (P + I) control action, when equilibrium is restored there remains no proportional action. This is because there is no longer any error.

## 8.7 DERIVATIVE CONTROL ACTION

Although integral control action removes the offset which is unavoidable with proportional action alone, it actually makes the possibility of instability greater, because it adds a further delay into the control loop. The possibility of instability is unavoidable in any feedback control loop because the effect of the disturbance on the measured value cannot be detected until after the delays are incurred. In order to partially overcome this problem and decrease the possibility of instability, an additional control action can be added which 'anticipates' the maximum magnitude of the disturbance by increasing the proportional action factor temporarily in proportion to the rate of change of the measured value. This has the effect of increasing the control action initially, but as the error between the set point and measured value decreases the proportional action reverts to its normal magnitude. Although originally referred to as 'anticipatory control action' this is now known as 'derivative action' because the mathematical expression for the rate of change of a variable is 'the derivative'.

Derivative control action is not in fact a third independent type of action, but a way of modifying proportional action so as to reduce the disadvantage of delays around the control loop. It cannot be used independently. Derivative action, like proportional and integral action is numerically defined by the 'derivative action factor' which is the rate of change of the *process* measured value which will produce control action of the same magnitude as the normal proportional action for any given error. In order to generate derivative control action it is obviously necessary to measure not only the value of the measured value, but also its rate of change. For this reason derivative control action is not possible with the simple 'local' type of mechanism described so far; its generation will be discussed later for more complex mechanisms.

## 8.8 THE CLOSED LOOP

Any process control regulating mechanism or system has three essential elements:

  (i)    Measurement system
 (ii)    Process system
(iii)    Control system

Note that each of these elements is in fact a system or rather a sub-system. From the diagram of such a control system (Fig. 8.11) it can be seen that these sub-systems are arranged in a loop and the system is usually referred to as a 'control loop' for this reason. Because the system forms a loop, information about the behaviour of the process variable (pressure, flow rate, temperature, etc.) can be fed back to the control system by the measurement system, and for this reason it is referred to as a 'feedback control loop'. The feedback control loop is, and probably always will be, the basic element of 'automatic process control'; it has already been stated that this is not the same thing as automated process control.

The closed loop includes the process as well as the control mechanism and measurement systems and it is essential, therefore, to consider how the process behaves before considering the behaviour of the closed loop as a whole.

Provided the process is stable, a given change in the position of the control valve (or other final control element) will, after elapse of a period referred to as the settling time, cause a change in the measured variable. The ratio of the magnitude of this change in process variable *steady state* value to the magnitude of the change in the control action which caused it is referred to as the open-loop process gain. Open loop

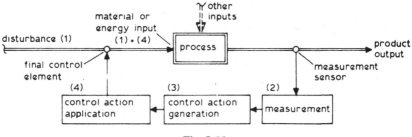

**Fig. 8.11.**

means that the controller is not operating because this is the only way in which such a ratio can be determined, since the purpose of the closed loop is to prevent change in the measured variable.

If there were no delays anywhere in the system, the settling time would be zero and, because any change in the measured variable is fed back and causes a control action which is in such a direction as to cancel it, there would be only a small change, dependent upon the proportional action factor. The greater the product of the process gain and the proportional action factor of the control mechanism the smaller this change; in fact with zero settling time there would be no limit to this factor, the change could be reduced to very small values, and there would be no need to employ either derivative or integral control actions, since proportional control would also be instantaneous in action. Needless to say, real systems always contain delays and therefore this scenario is totally unrealistic; it is also clear that the delays in any system are *entirely* responsible for the problems of control. It is therefore essential to start by considering the nature of these delays and how they arise. For the most part the delays in the process system are beyond the control of the designer of the control system (this subject will be dealt with later), unlike the delays in the control system or mechanism, and it is therefore even more important to understand how they arise and how they affect the design of a control system.

## 8.9 CAUSES OF DELAY IN PROCESS SYSTEMS

Whenever material (solid, liquid or gas) or energy (heat) flows into or out of a vessel, it takes time — instantaneous transfer is contrary to the laws of physics. Thus, the level of a liquid, the pressure of a gas or the temperature of the vessel contents cannot change suddenly but are subject to delays which are dependent upon the magnitude of the capacity and the resistance to the inflow of material or energy. Such delays are referred to as capacitive/resistive delays for this reason. It is a feature of such delays that the force causing transfer diminishes as equilibrium is approached, as, for instance, where the temperature of a vessel's contents approaches that of the water or steam in the heating coil. For example, consider a system in which gas at a higher pressure than that inside a vessel flows into the vessel; as it does so there will be an increase of gas pressure in the vessel (Fig. 8.12). Pressure in the vessel increases until it is equal to the pressure which is forcing the gas in and

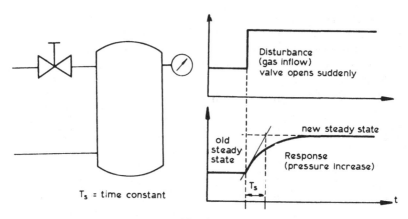

**Fig. 8.12.**

at that point the pressure ceases to rise — equilibrium, or 'steady state', has been reached and no more gas will enter.

This form of response to a sudden change of flow rate, supply pressure or heating (capacitive/resistive delay) is referred to in mathematical terms as 'exponential' delay. Note that the form of the response is the same whether pressure, flow rate or temperature is the measured variable. Such delay occurs in the process whenever a vessel has to be filled with liquid or emptied, or the gas pressure increased, or the temperature of its contents raised or lowered. Similarly such delays occur in measurement when heat has to be transferred through the wall of a thermowell into a filled system or thermocouple. Exactly the same type of delay occurs when the compressed air of the pneumatic control signal is increased in pressure and flows into the large capacity of the motor of a control valve. Hence, delays of this sort are encountered in the process, measurement and the application (rather than the generation) of control action.

Delays are also caused by the time taken for material to travel along pipes or conveyors from one place to another. This type of delay, shown in Fig. 8.13, is called 'dead time'.

All delays in either process or measurement are of one of these two types. Often, however, several delays are combined 'in series' (one after another) and when this happens it appears that there is just a single capacitive/resistive delay combined with dead time; Fig. 8.14 shows how this happens. Because of this it is always possible to consider *any* system as a single 'lumped' capacitive/resistive delay plus dead time.

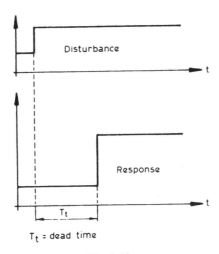

**Fig. 8.13.**

Only when the measurement variable *begins* to change in response to a change in the disturbance variable is it possible to take *any* control action. On the other hand the speed at which the control action must act on the process is proportional to the lumped time constant. It can therefore be seen that the difficulty of control is proportional to the ratio of dead time to lumped time constant, $T_u/T_g$ (Fig. 8.15).

## 8.10 DYNAMIC STABILITY

As was stated in Section 8.8, delays in the system are the cause of most of the problems of designing a control system: so far the response of plant items to one type of disturbance only has been considered — a step change in measured value. In fact this form of disturbance is very rare and it is essential that the response of units to more realistic types of disturbance is considered. The diagrams in Fig. 8.16 show the typical response to three types of disturbance. In practice disturbances will be some combination of these three types in most cases.

Because the control system, process and measurement system are connected in a loop, there is always the possibility of cyclical response and it is this which determines whether the system is stable or not. All cyclical responses can be considered as a combination of several sinusoidal responses of different frequencies and amplitudes; the effect

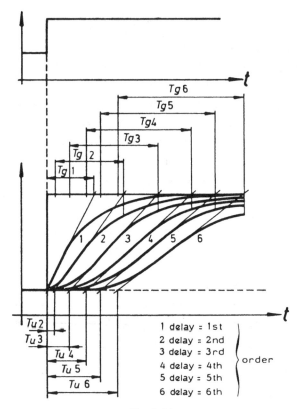

Fig. 8.14.

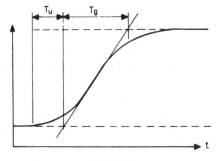

Fig. 8.15.

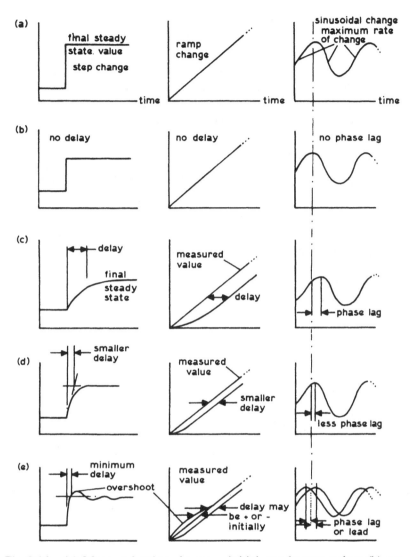

**Fig. 8.16.** (a) Measured value change — initial steady-state value; (b) proportional control action — can be generated without significant delays (hence it does not add to process and measurement delays); (c) integral control action — introduces delay or phase lag to any system in addition to measurement and process delays and lags; (d) proportional plus integral control action — introduces less delay or phase lag than integral action alone; overcomes 'offset' problem; (e) proportional plus integral plus derivative control action — may cause excessive control action to be generated if the measured value changes quickly, causing overshoot or phase advance. Derivative control action is difficult to use unless the *rate* of change can be reliably predicted.

of delays on sinusoidal response is to produce a phase lag, that is to say, the response at the measurement point 'lags' behind the input of the process. Once cyclical behaviour has set in it must be appreciated that the control action will also be of a cyclical nature; it should of course be in anti-phase to the disturbance so as to cancel it out, but because of delays in the process and measurement systems it will not be. If the delay is such that the control action lags the disturbance (or any given sinusoidal component of it) by half a cycle, or 180°, the control action will in fact be in phase with the disturbance instead of in anti-phase and, in respect of that particular sinusoidal component of the disturbance, the closed-loop system will be unstable if the product of the open-loop process gain and the control system action factor is more than unity. The higher the frequency of oscillation of any component sinusoid in the response, the more it will be attenuated in its passage through the process (by the natural damping of the process) and so the process will only be unstable if the frequency of the sinusoid, which will be delayed by half a cycle, is not in fact attenuated in its passage round the loop.

## 8.11 DESIGNING FOR DYNAMIC STABILITY

The criterion for dynamic stability of the *closed-loop* system — process plus measurement plus control — is that a sinusoidal disturbance of a frequency at which the delays impose 180° phase lag must be attenuated rather than amplified in its passage through the *open-loop* system — process and control (note that the latter includes the final control element). Assuming that such data are available either from design or test, then it can be plotted as a polar graph as shown in Fig. 8.17. At very low frequencies of disturbance there will be very little delay and very little attenuation; at zero frequency there will be none: hence the phase angle is zero and the magnitude of response a maximum. The magnitude will depend upon the process open-loop gain and the control action factor alone at very low frequencies, but at higher frequencies will depend also upon the attenuation introduced by the process response.

It will be recalled that, in addition to delays due to the process and measurement, the controller itself will cause delay if integral control action is used, whilst derivative control action achieves the opposite — phase advance. There will also be delays due to the operation of the final control element.

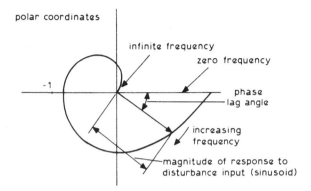

**Fig. 8.17.**   Phase/gain or Nyquist plot.

It is easy to see how the stability of an existing system can be established by carrying out a series of tests to find the phase lag and magnitude ratio when the system is disturbed by 'forcing' it with sinusoidal variations of varying frequency through the final control element. It is not easy to see how the system may be designed so that it is stable in the face of disturbances of a given speed (frequency). It is necessary, in order to do this, to be able to calculate the effect of the process delays on the closed loop response.

Figure 8.18 shows a typical capacitive/resistive process element; in this case liquid flows into a vessel and out through a restrictor valve (the resistance). The outflow depends upon the head of liquid in the vessel and the resistance of the valve (and other pipework). Any change in the inflow rate will cause a mismatch between the inflow and the outflow, which will not immediately change. Because of this mismatch the level in the vessel will begin to rise or fall and the head (and therefore the outflow) will start to change. As the head approaches the value which will again give equilibrium between the inflow and outflow (the new

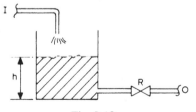

**Fig. 8.18.**

steady state), the mismatch will diminish and the rate at which the head changes will slow down, giving the types of response described earlier which is typical of any capacitive/resistive process element.

The 'dimensions' of the system are defined as the capacitance of the tank, $C$, and the resistance of the valve, $R$. Capacitance is defined as the rate at which the head in the vessel (the measured value) increases or decreases for unit increase in the inflow rate; resistance is defined as the head required to support an outflow of unit dimension (any consistent set of units can be used). The rate of change of head *at any instant* is obviously proportional to the mismatch between the rate of inflow $I$ and outflow $O$ and also to the capacitance. It can therefore be expressed mathematically as

$$\frac{dh}{dt} = \frac{1}{C}(I - O) \qquad \text{(assuming for simplicity that flow is laminar)}$$

but

$$O = h/R$$

hence

$$\frac{dh}{dt} = \frac{1}{C}\left(I - \frac{h}{R}\right)$$

or in operator form

$$\frac{d}{dt}h = p \cdot h = \frac{1}{C}\left(I - \frac{h}{R}\right)$$

hence

$$p \cdot h = \frac{I}{C} - \frac{h}{CR} \quad \text{or} \quad h\left(p + \frac{1}{CR}\right) = \frac{I}{C} \quad \text{or} \quad \frac{h}{I} = \frac{1/C}{\left(p + \frac{1}{CR}\right)}$$

The operator p is used to represent purely transient behaviour, and in order that the expression above shall represent both transient and steady state behaviour, p is replaced by s, the Laplace operator.

Now the dimensions of $C$ are

$$\frac{\text{length}^3}{\text{length}} = \text{length}^2$$

whilst the dimensions of $R$ are

$$\frac{\text{length}}{\text{length}^3/\text{time}} = \frac{\text{time}}{\text{length}^2}$$

Hence the dimensions of $CR$ are time $[T]$ and CR is in fact the time constant of the process unit. Thus, the relationship between the measured value $h$ and the disturbing variable $I$ is given by:

$$\frac{h}{I} = \frac{1/C}{\left(s + \dfrac{1}{T}\right)}$$

which is known as the transfer function of the process plant unit, and is usually denoted $G_{(s)}$.

A process element with two capacitive/resistive delays will have a transfer function (i.e. an expression in Laplace operator form defining the steady state and transient relationships between the measured variable and the disturbing variable) of the following form:

$$G_{(s)} = \frac{1/C_1 C_2}{\left(s + \dfrac{1}{T_1}\right)\left(s + \dfrac{1}{T_2}\right)} \quad \text{or} \quad \frac{T_1 T_2/C_1 C_2}{(sT_1 + 1)(sT_2 + 1)}$$

The denominator of this, the 'characteristic expression', can be expanded. The characteristic equation is formed by setting this expression equal to zero:

$$s^2 + (1/T_1 + 1/T_2)s + 1/(T_1 \cdot T_2) = 0$$

solving for s:

$$s = \frac{-(1/T_1 + 1/T_2) +/- \sqrt{(1/T_1 + 1/T_2)^2 - 4/(T_1 \cdot T_2)}}{2}$$

Now, if the value of the term under the square root sign is plotted for a range of values of the ratio of $T_1/T_2$ (Fig. 8.19), it is found that, whilst it is

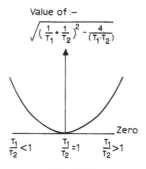

Value of :–

$$\sqrt{(\tfrac{1}{T_1}+\tfrac{1}{T_2})^2-\tfrac{4}{(T_1 \cdot T_2)}}$$

Zero

$\dfrac{T_1}{T_2}<1 \qquad \dfrac{T_1}{T_2}=1 \qquad \dfrac{T_1}{T_2}>1$

**Fig. 8.19.**

always positive, its minimum value is zero when this ratio is unity. In other words, when the two time constants are equal the system, although it does not actually display oscillatory behaviour, is on the point of doing so. The more unequal the two time constants are, the further the system is from the point at which it would begin to oscillate. So it can be seen that the term under the square root sign will *always* be positive or at worst zero, the roots of the 'characteristic expression' will always be totally 'real' and the system will not oscillate just so long as no feedback path exists within this second-order (two time constant) system!

A block diagram can be drawn of a simple plant/control mechanism which incorporates the transfer functions of the plant and the control mechanism (Fig. 8.20). The input to the plant is

$$E_{(s)} = (i - o)_{(s)} - H_{(s)} \cdot d_{(s)}$$

but

$$\frac{d_{(s)}}{E_{(s)}} = \frac{1/C}{(s + 1/T)} = G_{(s)}$$

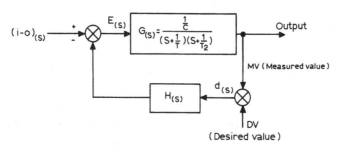

**Fig. 8.20.**

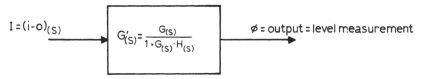

**Fig. 8.21.**

Thus, substituting for $E_{(s)}$:

$$d_{(s)} = G_{(s)} \cdot \{(I - O)_{(s)} - H_{(s)} \cdot d_{(s)}\}$$

and hence

$$\frac{d_{(s)}}{(i - o)_{(s)}} = \frac{G_{(s)}}{1 + G_{(s)} \cdot H_{(s)}}$$

Thus the block diagram, Fig. 8.20, can conveniently be replaced by that shown in Fig. 8.21, which contains the transfer function of the closed-loop system.

The 'characteristic function' of the closed loop transfer function can be seen to be

$$s^2 + \{1/T_1 + 1/T_2\}s + \{1/(T_1 \cdot T_2) + H_{(s)} \cdot /(C_1 \cdot C_2)\}$$

or

$$s^2 + \{1/T_1 + 1/T_2\}s + \{1/(T_1 \cdot T_2) + K \cdot /(C_1 \cdot C_2)\}$$

if the control mechanism in Fig. 8.19 can be assumed to have proportional action only and to operate instantaneously for all practical purposes.

The presence of the constant term

$$K \cdot /C_1 \cdot C_2$$

in the expression for s which results this time from solving for s, makes it possible for the expression under the square root sign to be negative, and thus for these roots to have 'imaginary' or oscillatory components. It is the addition of a feedback path within the closed-loop system that has introduced the possibility of oscillatory behaviour into a system which, without it, could not display transitory or oscillatory behaviour. Moreover it can easily be shown that this is true whether the two time-constant elements are both in the plant, both in the measurement/control mechanisms or one in each.

Dead time can be added, making the transfer function of a second order system with dead time

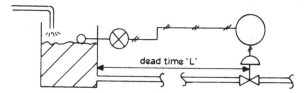

**Fig. 8.22.**

$$\frac{k \exp(-Ls)}{(sT_2 + 1)(sT_2 + 1)}$$

where $k = 1/C_1 C_2$. The additional phase lag is given by $L \times f \times 360°$ where $L$ is measured in appropriate units of time, and where $f$ = frequency. There is of course no attenuation associated with dead time, which is why it is so troublesome.

If a level controller were added to the vessel shown in Fig. 8.18 the 'motor' of the control valve would constitute a capacitive/resistive delay and there might well be significant dead time between the vessel and the control valve (Fig. 8.22). The transfer function through the process, measurement and control system might then approximate to

$$\frac{k \exp(-Ls)}{(sT_1 + 1)(sT_2 + 1)} = G_{(s)} H_{(s)}$$

The maximum phase lag that can result from any one capacitive/resistive delay is 90°, and so the Nyquist diagram for a system such as the open-loop (without control) system above will be of the form given in Fig. 8.23(a). Two such delays give a maximum of 180° phase lag and the Nyquist diagram will be of the form given in Fig. 8.23(b).

It can be seen that even with two such delays the magnitude of response will always be less than unity (i.e. it will be attenuated) however large the control action factor, and therefore the system can never be

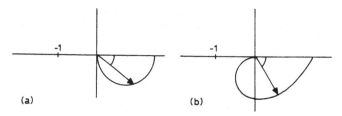

(a)                    (b)

**Fig. 8.23.**

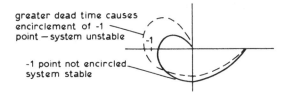

**Fig. 8.24.**

unstable though it may well oscillate. However, the addition of dead time increases the phase lag without any further attenuation, and instability becomes possible. The Nyquist diagram will then be as given in Fig. 8.24; if the −1 point is 'encircled' meaning that the response is actually amplified rather than attenuated, for a sinusoidal disturbance of such a frequency that the phase lag is 180°, the system is unstable in closed loop.

## 8.12 THE EFFECT OF INTEGRAL CONTROL ACTION ON DYNAMIC BEHAVIOUR

In the previous section it was shown that the addition of a feedback path to a plant which consisted of two dominant time-constant elements will introduce the possibility of that system displaying oscillatory response to a stimulus, particularly if the two time constants are equal or nearly equal. The simple plant/control mechanism shown in Fig. 8.25 will not oscillate, even though there is a feedback path through the control mechanism. This is because the process is first-order and the control,

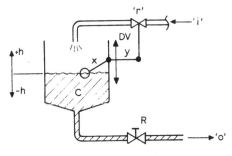

**Fig. 8.25.**

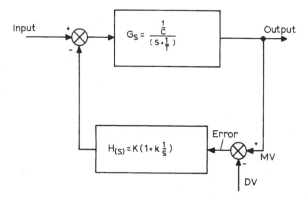

**Fig. 8.26.**

with proportional action only, can be assumed to act instantaneously so that the closed loop system is also first-order.

Since the operator s represents the derivative, $1/s$ must represent the integral, and the transfer function of a controller having both proportional and integral (auto-reset) control actions must be

$$K(1 + k_i/s)$$

where $K$ is the proportionality constant and $k_i$ the integral action constant.

Thus the closed-loop block diagram for a first-order plant with P + I control is as shown in Fig. 8.26. The closed-loop transfer function is thus

$$\frac{G_{(s)}}{1 + G_{(s)} \cdot H_{(s)}} = \frac{\dfrac{1/C}{(s + 1/T)}}{1 + \dfrac{1/C}{(s + 1/T)} \cdot K(1 + k_i/s)}$$

$$= \frac{1/C \cdot s}{s^2 + (1/T + K/C)s + K \cdot k_i/C}$$

and the characteristic function, the denominator of the transfer function, is second-order as a result of the inclusion of the integral, in addition to proportional control action in the feedback path. Thus, as has already been stated in Section 8.7, the addition of integral control action has the same effect on the propensity for a first-order plant to oscillate as the addition of a second time-constant element to the plant. On the other hand the steady-state error (the residual change in output which remains after the transient response to any disturbance has died away) is eliminated by the addition of integral (or automatic reset) control action. This is easily demonstrated by setting s, the rate of change, equal to zero in the transfer function which represents the ratio of response to disturbance. From the above it can be seen that this yields a zero value for the residual output — no steady-state error.

## 8.13 DAMPING FACTOR

The ratio of subsequent over- and undershoot of the measured process variable during the transient phase of its response to a disturbance, is a measure of the 'damping' in the system. If such cyclical over- and undershooting were to continue indefinitely, like the swing of a clock pendulum, then the damping ratio would be zero: again if no over- or undershoot at all occurs, as is the case with a first-order system, then the damping ratio is greater than unity. Between these limit conditions a damping factor of greater than zero but less then unity is a measure of the endurance of cyclical response to any disturbance which may be inflicted on the system. This is of considerable importance in process control since, if this period of unsteady behaviour is too prolonged, the plant will in practice never settle down at all. On the other hand, if there is too much damping in the system then, although no oscillation will occur, the plant will only reach a new steady state after an inordinately long period. The 'natural frequency' at which the plant would 'cycle' in the absence of any damping at all, and the damping ratio, must both be taken into account in assessing the system's behaviour. If each cycle takes 12 hours to complete, which is quite commonly true in refining systems or building heating systems, then a high value (close to unity) of damping factor is essential. On the other hand, too high a damping factor (much greater than unity) means that such a 'slow' system will take too long to attain a new steady state, which may well be just as bad as if the system went on oscillating for a long time!

It can be shown that the coefficient of the second term of the characteristic function of a second-order system is closely related to the damping factor, whilst the coefficient of the third term is the square of the natural frequency at which the system will cycle if the damping factor is reduced to zero. The characteristic function of a first-order plant controlled by a $P + I + D$ (proportional plus integral plus derivative) 3-term controller yields a damping factor

$$\frac{(1/T + K/C)}{2 \cdot (K \cdot k_i \cdot 1/C)^{1/2}} \cdot \frac{1}{(K \cdot k_d \cdot 1/C + 1)}$$

from which it can be seen that increasing either the integral action factor $k_i$ or the derivative action factor $k_d$ has the effect of reducing the damping factor and making the closed-loop system behaviour more 'lively'. Tuning of a 3-term SISO controller, by adjusting the three control action factors $K$, $k_i$ and $k_d$, in order to obtain optimal system dynamic behaviour, is obviously a highly skilled affair!

The behaviour of higher-order systems (containing more than two significant time constant elements) is much more complex. However, such systems are quite rare in practice; as shown in Section 8.9 multiple time-constant elements can be shown to behave as though there were one 'lumped' time constant plus dead time. Such systems, of which multiple stage distillation is perhaps the best example, are very difficult indeed to control (see Section 10.12).

## 8.14 THE SISO PROCESS CONTROLLER

The examples taken in earlier sections to illustrate the principles of control action have been of simple 'self-acting' mechanisms, such as were devised to make possible the regulation of process parameters *at local level*. Modern plant demands much more sophisticated mechanisms; in most cases these accept transmitted signals representing measured value and transmit a control signal over quite large distances to a final control element. The modern SISO (single input/single output) process controller generates control action which is a combination of proportional, integral and derivative actions, but it is much more flexible and can be combined with the other controllers in several ways, as will be seen, to enable complete control systems to be designed for the largest plant.

SISO controllers require a power supply (the motive force for a self-acting controller can be derived from the process pressures), which can be pneumatic, hydraulic or electrical. In diagram form the control loop using a SISO controller can be depicted as shown in Fig. 8.27. From this diagram it can be seen that the process and the control and measurement systems can be regarded as a system. Material and/or energy enters this system in two places and leaves as product. This type of system is called an 'open' system because energy and material enter

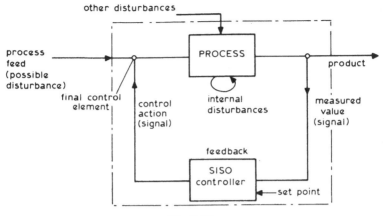

**Fig. 8.27.**

and leave it (a closed system is one where nothing enters or leaves).

A practical example of the above system is given in Fig. 8.28, which shows a heat exchanger in which heat is extracted from waste steam to raise the temperature of the process material before it enters some other processing stage. Disturbances can occur in the flow rate of the 'product feed' which will cause the temperature of the 'product to tankage' to change. Other disturbances can arise owing to change of pressure or flow rate of the waste steam. If the condensed steam is not immediately removed from the heat exchanger coils it may 'back up' and reduce the heat transfer, causing internal disturbances.

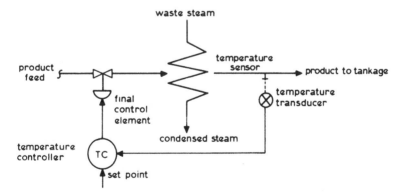

**Fig. 8.28.**

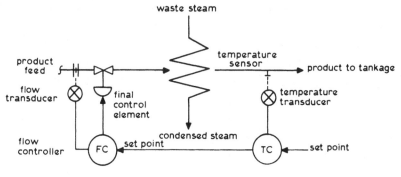

**Fig. 8.29.**

## 8.15 CONTROL SYSTEM DESIGN

The objective of control is to suppress any disturbance as early as possible: variations in the flow rate of the product feed in the system shown in Fig. 8.28 will disturb the measured value, i.e. product temperature, but only after delays due to heat transfer in the heat exchanger. It has been seen that these delays make feedback control very difficult. If the variations in flow rate are themselves controlled, there will be no disturbance in product temperature due to this cause. Another SISO controller can be added to the system to control the flow rate (Fig. 8.29). However, temperature will still be disturbed by other causes and so the temperature controller must cause the (controlled) flow rate to vary in order to compensate for these other disturbances. This use of one controller to 'reset' another is called 'cascade control'.

The delays in measurement of flow rate and in the flow itself (process) are very small compared to those of heat transfer, and very fast control of flow is possible (large proportional action factor). Quite fast disturbances of temperature, which could not be controlled by a single controller, can be controlled by a cascade system.

If the pressure of the waste steam rises or falls, the rate of heat transfer will rise or fall with it, especially if the temperature difference between the steam and the process material is small. Thus, changes in steam pressure will also disturb the product temperature and, since it is waste steam, it may not be possible to control this pressure. If, however, the steam pressure is measured (relative to a datum equal to the pressure at which the steam temperature equals the set point of the product temperature controller — suppressed zero) and the product flow rate

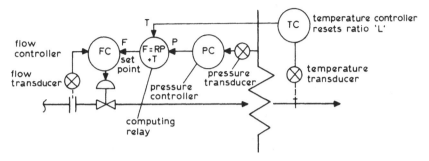

**Fig. 8.30.**

reset in 'ratio' to it, little disturbance to the product temperature will be caused by changes in steam pressure. By adding feedback control action generated by the temperature controller, the process will suffer minimum disturbance from either change of product flow rate or steam pressure (Fig. 8.30). There is no way that the effect of disturbances arising internally in the process, such as condensed steam back up, can be reduced by control — this is a matter for better plant design. The above description of a system design illustrates the way that more than one SISO controller can be used to form a complete control system. There are, however, difficulties associated with these techniques which will be discussed in Section 8.18.

## 8.16 PROPORTIONAL BANDWIDTH

The proportional action factor — the ratio of the control action taken to the actuating or measuring movement — is not applicable to a SISO controller that receives an analog signal which may represent any measuring range, and that generates a control *signal* which, when applied to the final control element, may produce any control action (depending upon the size of the final control element and the process characteristics). The term used to express the ratio of the control action *signal* generated by proportional action to the measurement error *signal* is 'proportional bandwidth'. Proportional control action is generated in a controller by amplifying the error (set point — measured value). The amplification factor is known as the 'gain' and the proportional bandwidth is the inverse of the gain expressed as a percentage. In other words:

$$\text{control action signal} = \text{error} \times \text{gain} \times -1$$

$$= \text{error} \times \frac{100}{\text{proportional bandwidth}} \times -1$$

Thus, with the controller 'tuned' to a proportional bandwidth of 100%, the maximum control signal will be generated when the error equals the full measurement range. That is to say that the maximum control action will not be applied to the process until the measured value has departed from the set point by an amount equal to the maximum which can be measured. For smaller errors the proportional control action will be less than the maximum possible. Of course, where derivative control action is added, the maximum control action may be applied for an error of much less than this but only whilst the measured value is actually changing. When integral control is added, the steady-state value of the control action can eventually reach any value within the analog range, however large the proportional bandwidth (however small the gain).

It can be seen from Sections 8.6 and 8.7 that the integral action factor and the derivative action factor both vary in proportion to the proportional action factor, and are defined as times: they are often referred to as the integral time and the derivative time. Tuning a SISO controller therefore consists of setting the proportional bandwidth, the integral time and the derivative time, so that the total control action best suits the behaviour of the process.

## 8.17 SELECTING THE CONTROL ACTION TO SUIT THE CONTROL SYSTEM

It was shown in Section 8.7 that the addition of integral control action makes it more likely that the controlled system will be unstable. However, without integral action there will always be a steady-state offset because an error is required in order to generate any control signal at all. Since the control signal generated by proportional action depends also on the gain, the offset will be greater for lower gains, i.e. for greater proportional bandwidth. How much gain can be used depends upon the effect of the process and measurement delays and, therefore, it may not be possible in controlling some processes with proportional control action alone to avoid large offset. Taking the example of the heat-exchanger system in Section 8.15, it would not be possible to avoid large offset in the case of the simple single temperature controller system because of the heat-transfer and measurement delays. However, the flow controller in the seconde cascaded system is not subject to either measurement or process delay of any size and could easily be tuned to a small proportional bandwidth, giving little offset. Moreover, the integral action of the temperature controller will reset the set point of the flow controller to a value which will make the measured value of temperature exactly equal to the set point. The control system is intended to control product temperature and so the offset in the flow controller is really unimportant. The temperature controller is called the master or primary controller, the flow controller the slave or secondary controller of the cascade system.

There is one circumstance when it might be necessary to use integral control action on a slave controller, that is if a small gain (large proportional bandwidth) has to be used. If the proportional bandwidth were greater than 100%, it would be impossible to generate maximum control action (the error would have to be greater than the measurement range to do so) using only proportional action and the slave controller would therefore not be able to respond to all the demands of the master controller unless it (the slave) had integral action. Fortunately, such conditions are rarely met, as slave controllers are usually capable of fast response (small proportional bandwidth).

## 8.18 FEEDFORWARD CONTROL

The system described in Section 8.15 (Fig. 8.30) uses three SISO controllers, two of them in a feedback cascade (Fig. 8.31). The third acts in a 'feedforward' fashion, and is not part of the feedback loop at all. Feedforward control is open-loop control, i.e. measurement, control and process elements do not form a loop at all, as can be seen from the diagram. This form of control cannot be used on its own, as there is no feedback of information to enable the steady-state value of the measured value to be compared with the set value. The aim of feedforward control action is to reduce the effect of disturbances on the measured value, by reducing the size of the disturbance itself before it enters the process. Feedback control is then applied to control the effect of the residual disturbance; it can never be 100% effective in this because

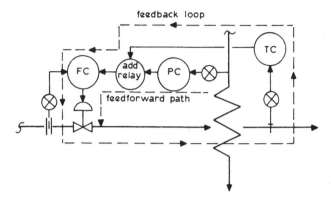

**Fig. 8.31.**

control action cannot be generated until the measured value has started to change in response to the disturbance.

The proportional action bandwidth of the pressure controller is adjusted so that the flow of process fluid is regulated in proportion to the rate of heat exchange, as this varies with steam pressure. Disturbance of the product temperature which would otherwise result from such steam pressure fluctuations will be greatly reduced; however, the effect of the changes of flow rate of the process fluid will not be fed back to the pressure controller and it can be seen, therefore, that this control action is feedforward.

Whereas delays are introduced into the feedback control loop by the process and measurement as described above, feedforward control action is likely to act too early. For instance, in the example of the heat exchanger system it will be some time after the steam pressure changes that the process fluid passing through the heat exchanger will start to vary in temperature; this is because heat is stored in the metal walls of the heat exchanger itself. Often suitable delays are introduced into the feedforward path in order to match these process delays and ensure, as far as possible, that the feedforward control action takes effect at the right time (something that cannot be achieved with feedback action). A 'first-order' delay can be achieved quite simply as shown in Fig. 8.32 for both pneumatic and electronic systems. The components required are a capacitor and a resistor (or needle valve in the case of pneumatics). The resistance is adjusted to provide the right time constant. Second- or higher-order delays can be used, enabling a process with small dead

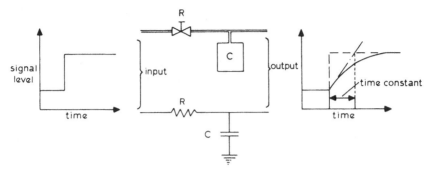

**Fig. 8.32.** First-order delay.

time to be matched, but dead time itself is almost impossible to achieve (see Section 8.9).

## 8.19 INTEGRAL SATURATION

Proportional control action, and therefore the position of the final control element, changes with the error between measured and set values. Integral action, however, continues to either increase or decrease as long as there is an error and, unless the error is removed, it will eventually drive the final control element to the physical limit of its travel. In controllers which add the two control actions, proportional and integral, together, either action alone must be capable of driving the final control element over its full range; hence, once integral control action has driven the control valve (final control element) to its wide open or closed position, it is not possible for the proportional control action to cause it to even start to move in the other direction as the measured value responds to full control action and the error begins to reduce. In fact, when, as a result of applying maximum control action, the measured value changes so that the error decreases to zero, as it must in due time, proportional action will be zero and the valve will still be firmly at its limit. Only when the error has changed sign and increased sufficiently and for a sufficient time will the combined proportional and integral control actions *begin* to move the valve away from its limit position. This state of affairs is made much worse by the fact that control action can in practical controllers exceed the analog limits (0·2 bar, 4 mA; or 1·0 bar, 20 mA), which results in the integral control action taking even longer to desaturate or 'unwind'.

The problem of integral saturation (windup) can occur with processes which have considerable dead time or very long 'lumped' time constants. It is an almost invariable problem with 'batch' or discontinuous processes because during periods when the process is not operating the measured value cannot equal the set point and it is inevitable that integral control action will saturate, unless precautions are taken to limit integral action to less than the full analog range.

## 8.20 MANUAL CONTROL AND 'BUMPLESS TRANSFER'

Under certain operating conditions or circumstances (for instance, start up, or for maintenance) automatic control action is stopped and the set point generating mechanism is used simply to position the final control element without reference to the measured value at all. This is 'open-loop control' and is usually referred to as 'manual mode'. When the controller is being used in this way it is very difficult to stop the generation of control actions even though the process no longer responds to them. In any case it is important to arrange that the output of the controller, which is the sum of the control actions, does not change suddenly when it is switched back into 'automatic mode'. When there is no error, the output of any controller is only the integral control action term, since there is no proportional action without an error and no derivative action if the error is not changing. Therefore, to ensure 'bumpless transfer' from manual mode operation to automatic mode, two things are necessary: (i) the set point must equal the measured value, and (ii) the integral action must be equal to the controller output.

In order to achieve these two things, the measured value signal is connected to the set-point system inside the controller (to the set-point bellows in the case of a pneumatic controller) and the output of the controller (generated by the set-point mechanism in manual mode) is connected to the integral action generating mechanism (to the integral bellows in the case of a pneumatic controller). Thus, when the controller is switched from manual mode operation to automatic mode, the set point is equal to the measured value, whatever the latter may be. There is, therefore, no error and the integral action is equal to the output of the controller which does not therefore change. After transfer the set-point mechanism can be set by the operator to any value he requires and the controller will vary the output to the final control element smoothly until this measured value is attained.

## 8.21 EXTERNAL RESET

It should be remembered that the purpose of integral control action is to remove the offset which is otherwise inevitable with proportional control. It is helpful in this context to consider that rate or derivative control action is achieved by increasing or decreasing the proportional gain temporarily, so that rate action cannot in fact be dissociated from proportional control whereas integral control action is entirely separate and can be used alone. In a cascade loop the output of the master controller is the desired value (set point) of the slave controller; hence, if the measured value of the slave is fed back to the master and used as the reset signal, it will be seen that the desired value of the master controller will be satisfied *without error* when the process is operating in steady state. In steady state there is no proportional control action or rate action term and the output of the master controller, i.e. the set point of the slave controller, is equal to the integral control action term, in this case the measured value of the slave controller. Only when an error develops in the master-controlled variable will a proportional action term develop in the master controller and the set point of the slave controller be different from the measured value. Then the slave controller will act to change the slave-controlled variable, continuing to do so until the error in the master controller is removed, when the set point of the slave will again be equal to the measured value. The system can be appreciated from the diagram in Fig. 8.33 which shows a typical cascade system, comprising a composition controller as master with a measured value derived from an on-stream analyser and a flow controller as slave. Generation of integral control action in the normal way inside the master controller would serve the same end but would be prone to integral 'windup' (see Section 8.19) which is avoided by this method.

## 8.22 DISCRETE CONTROL ACTIONS

It has been assumed so far that the control actions generated are continuous, and position the final control element smoothly in a continuous manner. This does not necessarily have to be the case; in a type of control action known as 'boundless' the final control element is positioned in a series of steps at fixed intervals of time. Proportional and derivative control actions are generated in the same way as for the

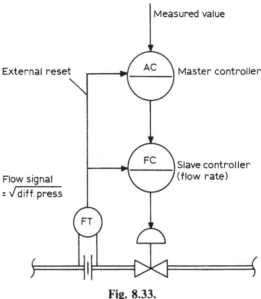

**Fig. 8.33.**

controllers so far described, i.e. from the error and rate of change signals. However, at the interval time the final control element is made to move an amount proportional to the combined control action signal in the appropriate direction. At the next interval time the control action will be different because the error will have changed, and the rate of change of the measured variable may be different. Again, the final control element will be made to change by an amount proportional to the control action. Each change will be *in addition* to the last and so the final control element is made to sum or integrate the control action. Since proportional action is proportional to error at all times, control action generated in this way is the same as for conventional control (apart from its 'jerky' application) but not if derivative action is added, since then the derivative component as well as the proportional component is 'integrated'. Such a controller will desaturate very quickly, both because the final control element (which is the integrator) cannot exceed the operating limits and also because the integration rate is increased by derivative control action, unlike the more conventional control action which simply adds the three 'terms', integral, proportional and derivative.

Some processes have such large 'lumped time constants' that control action can only cause the measured value to change extremely slowly (e.g. space heating). In these cases a very simple type of controller is possible which generates 'proportional time control action'. A suitable time interval is selected and within this time the final control element is set to maximum for a period and then to minimum for the rest of the interval. The ratio of the time which the final control element (which is often a simple switch) spends in the 'maximum' position to the time it spends in the minimum position is proportioned to the error. Thus, over a relatively long period of time the average control action is proportional to the error. Integral and control action terms can be generated in the same way as for other controllers, the ratio of times at maximum and minimum being equivalent to the position of a 'modulating' final control element.

## 8.23  DIRECT DIGITAL CONTROL

Direct digital control (DDC) is a term which has come to be used to describe control action generated by digital rather than analog computation, as, for instance, by a digital computer. The control action is exactly the same as has been described in the previous sections but it can be convenient to transmit it to the final control element in digital form to drive a stepper motor, for instance. Generally digital computation is more accurate than analog and it is possible to switch integral action out in a digital controller, which effectively solves the saturation problem. There are other advantages and disadvantages of DDC, but these will not be discussed here.

# CHAPTER 9

# *Final Control Elements*

## 9.1 INTRODUCTION

The final control element applies the control action, generated in the controller mechanism, to the process; if it does not perform its function adequately there will have been little point in measuring or in generating control action. Hence, the final control element is in some ways the most important part of the control system. Certainly it is often given too little attention, with the result that the control system performs very badly.

Processing is usually controlled by regulating the flow of gases, vapours, liquids or fluidised powders through pipelines. For this reason, the final control element is almost always a throttling valve; however, it may be a thyristor control element for regulating the flow of electrical energy to an electrical machine, a damper in a ventilation or forced draught duct or any regulating mechanism. Whatever it is, it will have a characteristic relationship between its input (the control action signal generated by the controller) and its output (the effect it has on the process), which will not usually be one of simple proportionality. In addition, its response is unlikely to be instantaneous — there will be delays as there are in both the process and the measurement. It should be appreciated, therefore, that the final control element will inevitably modify the control action generated in such a way that control will be different under different processing conditions. As in the case of the process and the controller, it is necessary to understand how the input/output relationship of a control valve affects the closed loop.

In order that a fluid can flow through a given piping system there must be a pressure difference between the input and the output of the system: in order that the flow rate can be adjusted, that pressure difference must be manipulated — reduced to reduce flow rate, increased to increase flow rate. Less power is required to drive the lower flow rate, and the function of the throttling control valve is to waste the additional power available in the system from a pump or other driving force provided to cater for the maximum flow rate required. It is in fact much less energy-wasteful to manipulate a speed control system on the pump as the final control element than to throttle, but the capital investment is much higher. In the case of a gas or vapour, it would always be less energy-wasteful to manipulate the pressure of generation (a steam boiler for instance), but this is not usually practicable. In the majority of systems, therefore (with the exception of those which use a very large amount of power, such as long distance pipelines), unneeded power is simply wasted by throttling flow through a valve. *A control valve, therefore, is a device for wasting energy so that the required measured value can be maintained.* In doing so it must convert the analog signal, which is the output of the controller, into a corresponding change in the process variable.

## 9.2 THE CONTROL VALVE

The control valve is simply a variable restrictor in the process pipeline; it can take many forms, as shown in Figs 9.1–9.3. Whatever form it takes however, it must have certain essential elements:

(i)   a means of varying the opening through the valve;
(ii)  a mechanical linkage to (iii);
(iii) an operator or motor, which translates the analog control signal into motion.

In the case of the globe valve (Fig. 9.1), (i) takes the form of a plug and seat, or for the 'double-beat' valve two plugs and two seats, (ii) is a stem and (iii) is a diaphragm 'motor'. In the case of the ball valve (Fig. 9.2), (i) takes the form of a ball with a large hole drilled through it, which seats against a seal ring. The butterfly valve (Fig. 9.3) uses a disc, pivoted about an axis as shown, so that in one position it presents its narrowest profile to the flow, whilst pivoted through 90° it completely blocks the pipe. Both the ball and butterfly valves are operated by an arm and spindle, which rotates in bearings, and the operator may also be a diaphragm motor or other pneumatically, electrically or hydraulically

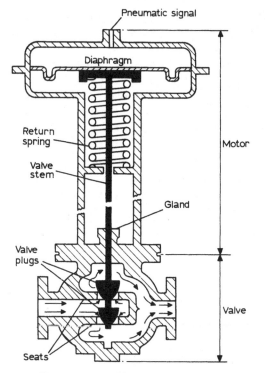

**Fig. 9.1.** Double-beat globe valve.

powered motor. All valves also have a gland at the point of exit of the mechanical linkage.

The operator or motor shown on the globe valve in Fig. 9.1 positions the stem: the pneumatic analog control signal is connected into the

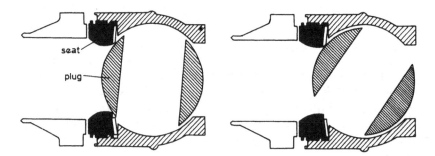

**Fig. 9.2.** Ball valve.

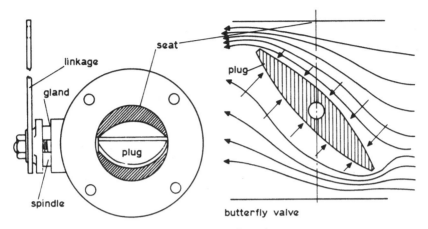

**Fig. 9.3.** Butterfly valve.

diaphragm chamber, and the pressure acting across the area of the diaphragm provides a force which is opposed by the spring and by whatever forces are produced by the process pressures acting on the valve plugs. As the analog signal pressure increases, the spring is compressed and the stem moves downwards closing the valve and 'throttling' the flow of process fluid through it. Since the spring force is proportional to the distance through which it is compressed, the movement of the valve stem is proportional to the change in the analog signal pressure. Two causes of error in this relationship are:

(i)  the forces acting on the plug of the valve, which vary with the process conditions and with the position of the valve (these are called the out-of-balance forces); and

(ii)  'stiction' between the stem and the gland, which is inevitable.

The 'double-beat' valve shown is really two valves mechanically linked. This is done to reduce the out-of-balance forces as much as possible. Flow is upwards through the top plug and downwards through the lower plug. Because there is a pressure loss through the valve (that is its function), the pressure acting upwards over the cross-sectional area of the plug produces a different force from that acting downwards. The resultant out-of-balance force is, however, upwards in one case and downwards in the other, and provided the cross-sectional areas are carefully balanced, the resultant out-of-balance force on the stem of the valve, which will produce error, is very small. Provided the spring force and the force produced by the diaphragm are much greater than this out-of-balance force, the error in positioning of the valve stem will be small. Some smaller globe-type control valves are 'single-beat', i.e. they have only one plug and seat. The out-of-balance forces in a single-beat valve will obviously be much greater than in a double-beat valve, and to ensure that this does not result in large errors of positioning, a larger motor and stronger spring are used. Nevertheless, single-beat valves cannot be used where there are large pressure drops through the valve, unless a positioner (see Section 9.3) is used to overcome these errors of positioning. Single-beat valves do have the advantage that they can control accurately at much smaller openings, because any inaccuracy of manufacture, damage or differential thermal expansion in a double-beat valve is likely to result in one plug 'seating' before the other.

The problem of out-of-balance forces is much worse in the case of the butterfly valve as can be seen from the graph in Fig. 9.4(a) which shows a plot of torque (rotary force) against the angle of opening of the valve. This shows clearly that the force required from the operator merely to hold the valve plug in position varies from almost nothing when the valve is slightly open to a very large maximum value when it is 80° open. A very considerable decrease in the out-of-balance torque can be achieved by using a 'fishtail' plug as shown in Fig. 9.4(b); the 'tail' breaks up the air flow in the same way as a 'spoiler' on a car and thus reduces the pressure acting on the face of the plug.

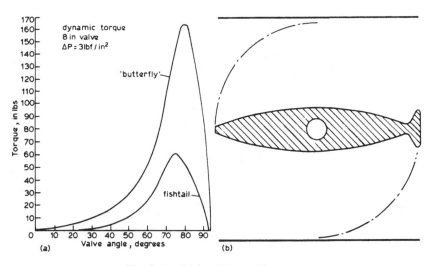

**Fig. 9.4.** 'Fishtail' butterfly valve.

The ball valve too has very variable out-of-balance torque characteristics which cannot be overcome.

## 9.3 POSITIONERS

The effect of out-of-balance forces (or torques), together with the static friction (or stiction) of the gland, introduce hysteresis and repeatability errors into the relationship between input signal and plug position. Where these errors would be unacceptable, the size of the operator can be increased so that its force and that of the spring are much greater than the error-producing forces. However, except in the case of the double-beat globe pattern valve, this may require an unacceptably large operator; in such cases a valve positioner can be fitted. The valve positioner is really a controller which has as its measured value the position of the stem of the valve, transmitted to it by mechanical linkage; it is mounted on the frame or 'yoke' of the valve itself. The control action is usually proportional only, so the errors are reduced but not eliminated altogether. The linkage can be 'characterised' (usually by using a cam) to change the relationship of the stem movement of the valve to the opening between plug and seat (the advantage of this will be seen later).

The positioner is in fact a 'slave' controller to the process controller, and the two are in cascade, as described in Section 8.15. In order that any tendency to oscillatory behaviour in the slave loop of a cascade system should not be transferred to the master controller (in this case from the positioner to the process controller) or *vice versa*, it is essential to observe the rule that the speed of response in the minor or slave 'loop' shall be at least four times faster than that in the major or master loop. (The 'interference' which would otherwise occur is similar to that which occurs between radio stations operating on 'frequencies' which are too close together.) It is rarely possible to keep this rule in the case of flow control, which is why positioners are not normally used on flow control valves.

## 9.4 CONTROL VALVE CHARACTERISTICS

The relationship between the movement of the stem of the valve, or other linkage, and the increase or decrease in the opening between the plug and seat, is not constant but varies with the valve position. The

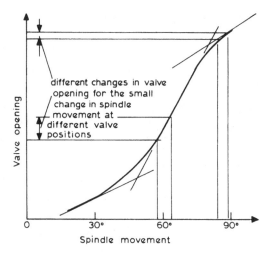

**Fig. 9.5.**  Butterfly valve characteristics.

valve characteristic is the name given to that relationship. The characteristic of an ordinary butterfly valve is shown in Fig. 9.5.

It can be seen that when the valve is nearly wide open or closed a given movement of the spindle causes the opening to change in area less than the same spindle movement does when the valve is near the middle of its movement. If the characteristic plot were a straight line, it can be seen that the change in valve opening for a given movement of the spindle would be constant, no matter what position the valve was in at the time. Such a characteristic is called 'linear'; it might be thought that this is the ideal characteristic, since valve opening will be proportional to input (the control signal) throughout the valve movement. However, two other characteristics are common and have considerable advantages in certain situations. These are 'quick-opening' and 'equal-percentage' characteristics (see Fig. 9.6).

The 'quick-opening' character is self-explanatory. The 'equal-percentage' character is such that at any point in the valve movement a given stem or spindle movement results in an equal fractional increase in the opening. Thus, when the valve is only a little open, the change in opening for the given stem or spindle movement is small, but when nearly wide open the change in opening for the same movement of the stem or spindle is much greater. One of the main advantages of this characteristic is that the valve positioning becomes progressively less critical as it nears the closed position, which effectively extends the

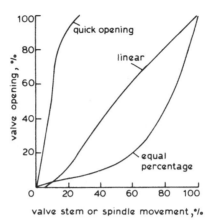

Fig. 9.6.

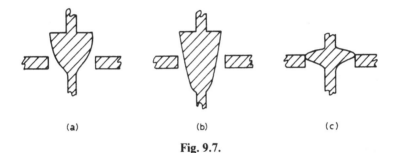

(a)      (b)      (c)

Fig. 9.7.

range of opening over which the valve can operate adequately. The butterfly valve characteristic is a mixture of these three characteristics and is therefore not likely to be very suitable for any application. The main advantage of the globe type of valve is that the plug shape can be made so as to give any characteristic (see Fig. 9.7).

The flow rate through the control valve depends upon the valve opening and the pressure difference between the inlet and the outlet — the pressure drop across the valve. It will also depend on such factors as the 'flow profile': for instance, the capacity of a right-angle-bodied valve will not be as large as that of a straight-through valve because the flow has to change direction in passing through the valve. On the other hand, a valve with a 'flared' exit will have a larger capacity. All valves have a capacity coefficient for any particular opening:

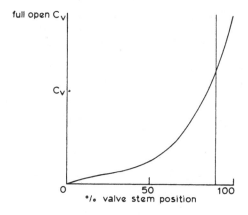

full open $C_V$

$C_V$

O        50        100

°/₀ valve stem position

**Fig. 9.8.**

$$C_v = f \sqrt{\frac{SG}{dP}}$$

where $f$ is the volumetric flow rate of the process liquid through the valve with a pressure drop of $dP$ across it, and $SG$ is the specific gravity of the liquid.

Any given valve can best be described therefore by its characteristic plot of $C_v$ against stem or spindle movement (Fig. 9.8).

## 9.5 INSTALLED CHARACTERISTICS

Because the flow rate through the valve depends on the pressure drop across it, it is not necessarily true that a linear characteristic is ideal. Ideally the change in the *process variable* resulting from a given change in valve stem or spindle position should always be the same, because only then will the control valve not modify the control action generated by the controller. However, in most systems the pressure drop across the valve changes with flow through the valve, and it is therefore not sufficient to ensure that the valve opening varies linearly with stem movement; the valve characteristic must be chosen so that it compensates for changes in pressure drop. This can be understood by reference to Figs 9.9 and 9.10.

In Fig. 9.9 the process system is such that the pressure drop across the valve reduces as the flow rate increases, until, when the valve is almost

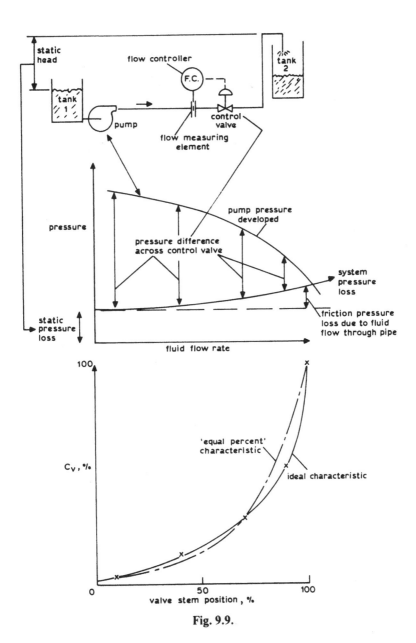

Fig. 9.9.

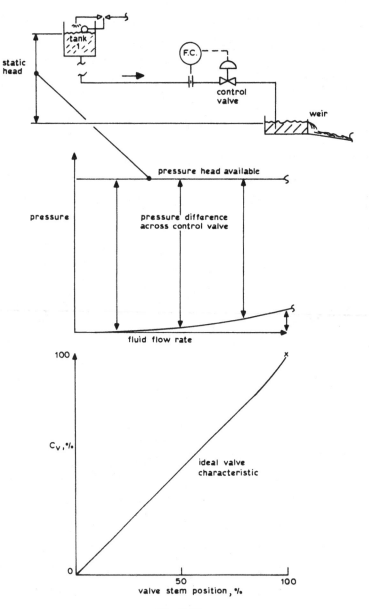

**Fig. 9.10.**

fully open, it is very small indeed. It is obvious, therefore, that the valve should open slowly when it is nearly closed, whilst it must increase its opening by a large amount for each unit movement of the stem as it nears the fully open position. By calculating the *required $C_v$* for different flow rates and plotting these in such a way that each increment in flow rate is proportional to the corresponding increment in valve stem movement (and hence to the control signal) as shown, the 'ideal' valve characteristic can be found. In this case it can be seen that it is very similar to the equal-percentage characteristic (Fig. 9.9). If a linear characteristic valve were used in this case, the proportional action factor would be very much smaller at high flow rates than it would be at low flows, and the system could not be tuned for the best response at both high and low flow rates.

In Fig. 9.10 the pressure across the valve is provided by static head rather than a pump, and although the pressure lost by friction in the pipe increases as flow increases, the pressure drop across the valve changes little between maximum and low flow rates. In this case, therefore, the flow rate will be approximately proportional to the valve opening, and the ideal characteristic can readily be seen to be close to the linear form. It would be just as wrong to install an equal-percentage valve in this system as it would be to install a linear valve in the first system.

The correct choice of valve characteristic will differ with *every* system and depends, as is seen from the examples in Figs 9.9 and 9.10, on the process characteristic as well as the valve characteristic. The ideal *installed* characteristic is that which makes the (steady state) change in the *process variable* proportional to the valve stem movement.

Often the ideal installed characteristic lies between equal-percentage and linear, and on other occasions it is more important to obtain the maximum rangeability, which favours the equal-percentage characteristic. This is why equal-percentage characteristic valves are chosen more often than linear, but *it must be remembered that if the characteristic does not suit the process, it will only be possible to tune the system for correct response under one set of operating conditions*. Often a compromise solution is acceptable, but in cases where both maximum rangeability *and* correct tuning are essential — as when the process system must operate over a wide range of operating conditions — the valve positioner offers a partial solution. As mentioned in Section 9.3 the linkage transmitting the valve stem position to the positioner can be characterised in such a way as to compensate for mismatch between process and valve characteristics. This is, however, a very difficult design problem and is only resorted to in exceptional cases.


## 9.6 VALVE SIZING

A control valve which is too large or too small cannot properly apply the control action generated by the controller from measurement of the process variable. The capacity of the valve at any opening is given by its $C_v$, and the values of this coefficient must be matched to the design normal, maximum and minimum requirements. The smallest flow rate which can be effectively controlled will obviously be related to the leak rate across the plug and seat of a fully closed valve. All valves leak, and the rate depends on the 'shut-off' pressure drop; the type of valve will also have a considerable influence on the leak rate. It was mentioned in Section 9.2 that differential thermal expansion may cause one of the plugs of a double-beat globe valve to seat before the other, which will obviously increase the leak rate considerably. In the butterfly valve the seating is very poor because of the difficulty of accurate guiding, and high leak rates are inevitable. On the other hand, a ball valve in good condition has a very low leak rate. Obviously the condition of the valve and the severity of service also, in practice, affect the leak rate and limit the lower end of the valve range. Taking all these factors into consideration, it is probably reasonable, and therefore good practice, to limit the operating range at the lower end to 10% of the full-open capacity in the case of double-beat globe valves and butterfly types, and 5% in the case of single-beat globe valves and ball valves. Even then it has to be remembered that, in order to control the process in the face of disturbances to the steady-state 'design' conditions, a control valve must be capable of increasing or decreasing the flow rate about a mean steady state *at all operating conditions.* Thus, in matching the valve capacity to the process, these limits become 20% and 10%; for the same reason the upper limit is usually taken as no more than 80% of the full-open value. Thus, the rangeability — the ratio of the maximum usable $C_v$ to the minimum usable $C_v$ — lies between 5:1 and 10:1, depending upon the type and condition of the valve. As was seen from the previous section, when the pressure drop across the valve rises as the flow rate falls, this range is reduced still further, and it is not uncommon to find that the *effective* rangeability is no more than 3:1 in practice.

One way in which errors of sizing can be corrected after commissioning of the plant is finished (and it can be seen whether the valve is *in fact* too large or too small) is to install a valve body with a 'reduced' size 'trim'. Most globe pattern valves can be obtained with a seat and plug one size smaller than that which the body of the valve would normally contain; thus, in effect, a smaller valve is installed with the option of increasing it later (it is much cheaper and easier to replace 'trim' than to remove the whole valve from the pipe once the plant has started up). However, this option is only possible in the case of globe-type valves (single- or double-beat), as can be appreciated from the construction of other types; this is one reason why this type of valve, though expensive, is still the most widely used. The other reasons are more precise positioning and greater flexibility of characterisation.

## 9.7 MAINTENANCE OF CONTROL VALVES

From the foregoing it must be obvious that the stem or spindles and the glands of control valves must be maintained in good condition at all times. Failure to do so will result in poor control usually long before it results in breakdown. Care should be taken to ensure that valves are installed in the pipeline in such a location (having due regard to control requirements, especially response) that access for maintenance is good. Often, in designing the pipework, it will be overlooked that sufficient clearance must be provided, not only to get at the valve but to withdraw its internal parts (trim, for instance) for inspection and repair without removing the body from the pipeline (which can usually only be done at a shut-down). This often prevents adequate maintenance, with disastrous results.

## 9.8 DYNAMIC RESPONSE

It was mentioned earlier in this chapter that, in addition to modifying the steady-state relationship between control action signal and process variable, the control valve introduces delays because of the time it takes for instrument supply air to 'fill' or 'empty' the large capacity of the operator or motor. This is likely to be a problem with 'fast' process loops, such as flow and some pressure controls, particularly where the controller is located in a central control room a long distance from the control valve (which is common today). A solution is often found by fitting an amplifier relay like that used in the controller (and usually referred to as a 'booster' relay) at the valve itself. This isolates the controller output signal from the valve motor, so that air from the controller, travelling through the considerable restriction of the signal line from the remote control room, has only to fill the very small capacity of the relay chamber, with the result that hardly any delay is generated in that part of the system. Air to fill the motor chamber is supplied at high pressure (5–7 bar) close to the valve and is subject to much less restriction; thus, the delay in filling the motor chamber is greatly reduced. If the delay is still too great, high pressure motors can be obtained, with the advantage that the same force can be generated with a smaller capacity, using air at a higher pressure.

## 9.9 CONTROL VALVES AS PLANT EQUIPMENT

The control valve is the only part of the control loop which is also an integral part of the plant itself (measurement sensors, which are also installed into the plant, do not take any part in operation). It is to be expected, therefore, that the proper specification of a control valve is as much concerned with process and mechanical considerations as with control.

A control valve must *never* be used as an isolating valve; it is not designed for this purpose, and its seats and plugs will probably be damaged to such an extent that control will be badly affected if it is used as a shut-off device. Where the plant operation requires them, isolating valves *must be installed in addition to control valves.*

The direction of flow through the valve is of great importance, except possibly in the case of a simple butterfly valve. The seal on a ball valve must face the flow; the tail of a fishtail butterfly must be downstream when the valve opens. In the case of a globe valve, both the operator and the capacity of the valve will be affected if the flow direction is changed. This is illustrated in Fig. 9.11, which represents a type of single-beat globe control valve. If flow is downward through the plug and seat, the upstream pressure, which must always be the higher, acts on the smaller cross-sectional area because pressure on the valve stem does not act downwards. Because of this, the out-of-balance forces can be reduced; in fact, it may be possible, if the stem is of a large enough diameter, to virtually eliminate them. Thus, the actuator can be smaller than would be required if the same valve were installed so that flow is upwards, in

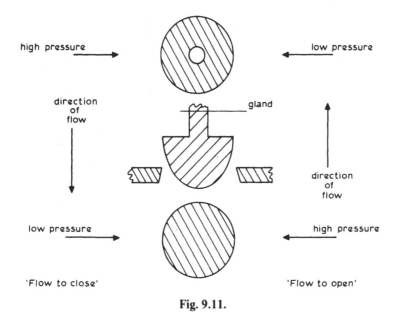

**Fig. 9.11.**

which case the highest pressure acts on the largest area, and it is not possible to reduce the out-of-balance forces.

Other considerations may dictate that flow must be in the direction which would maximise the out-of-balance forces on a single-beat valve, and this may be a reason for selecting a double-beat valve despite its greater cost and leak potential. Because the flow in a double-beat valve splits between the two 'plugs' there need be no out-of-balance forces, provided the valve is designed with the correct cross-sectional areas as shown in Fig. 9.12. For instance, it may be vital that if the pneumatic air supply fails the valve should open.

The capacity of a control valve may vary considerably according to the direction of flow through it, especially if the body has been designed to give a high 'recovery' on the downstream side by shaping it like a measuring nozzle or venturi (see Section 4.2).

In considering the mode of failure of the valve it should be remembered that the plug is only driven in one direction by pneumatic force; the spring provides the force to drive it in the other direction. Apart from the direction of flow through the valve, there is another option — whether spring or pneumatic pressure is employed to open or close the valve if there is a failure of the pneumatic supply pressure. Finally, it may be important to consider what will happen if the control

'Single beat' valves                    'Double beat' valves

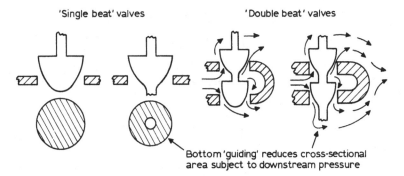

Bottom 'guiding' reduces cross-sectional
area subject to downstream pressure

**Fig. 9.12.**

signal should fail but not the pneumatic air supply (if electronic control
and transmission are used or if either a positioner or 'booster' relay is
used at the valve). The answers to all these questions will decide the size
of actuator required.

The process conditions under which the valve has to operate will
decide the type of construction appropriate to the internals of the valve.
For instance, the stem may be top guided only or may be supported top
and bottom if turbulence or 'flashing' (sudden change from liquid to
vapour as the pressure drops through the valve) is expected. The effect of
these options on out-of-balance forces and capacity, can be appreciated
from the diagrams shown in Fig. 9.12.

Materials of construction of the body, and particularly of the seat and
plug, are of considerable importance. One reason why it is not desirable
for the valve to be oversized is that it would then be operating with a
small clearance between the plug and seat for much of its life; this will
inevitably accelerate wear by erosion, particularly where the process
fluid contains solid matter (dirt) or where flashing may occur, and this
in turn will adversely affect control.

## 9.10 SPLIT-RANGE OPERATION

It was explained in Section 9.6 that the useful range of operation of any
control valve is often quite small. Fortunately, most process plant tends
to be operated close to design conditions; it is not uncommon, however,
to find that the rangeability available, when all the constraints have
been allowed for in the design and selection of equipment, is not

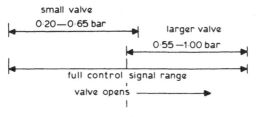

**Fig. 9.13.**

adequate. In such cases it is possible to install two control valves in parallel, one normally being larger than the other, to greatly extend the operating range. Each valve must have a positioner, which can be adjusted so that it operates over its full range of opening in response to only a part of the control signal range (Fig. 9.13). The operating ranges of the two valves overlap to ensure that there cannot be a part of the total operating range where no increase in capacity occurs for an increase in control signal; however, selection of the most appropriate valve characteristic can pose a considerable problem. In very special cases it would be possible to characterise both valve positioners so that the combined character of the two valves would be close to ideal. This is so difficult in practice that it is rarely if ever done, and increasing rangeability is normally achieved at the expense of installed characteristic and, therefore, poor tuning over the operating range.

## 9.11 MOTOR SPEED CONTROL

While it is true that the great majority of final control elements in process systems are throttling valves, there is increasing use of variable speed control of drive motors for pumps, fans compressors, etc., as an alternative to throttling away unwanted energy. In terms of the other parameters of the pump or compressor — flow and pressure — there is a different operating characteristic for each value of speed. Thus, the operating characteristic of the pump, fan or compressor can be altered so that the combined plant operating characteristic suits the process requirements without the addition of a throttling valve to the system. For instance, as shown in Fig. 9.14, if fluid flow rate is to be reduced, the operating point can be moved to the left in the system diagram by reducing the speed until the pressure produced by the pump exactly equals that dissipated in the system pipework, etc., at the lower flow rate.

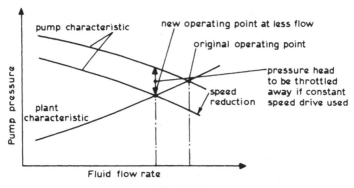

**Fig. 9.14.**

Variable speed drives have, in the past, consisted of DC motors: speed is varied in simple cases by means of a variable resistor in the armature circuit, manipulation of which varies the armature voltage and thus the motor speed. However, this is in effect a form of throttling, as energy is dissipated in heat to waste in the resistor which has to carry the full armature current. For large drives the Ward–Leonard system has been used; in this a separate motor/generator set is used to supply the armature current, but this is very expensive in capital terms and hard to justify economically for any but the largest energy users. In any case DC electric power is very expensive to distribute and would normally have to be generated on site or obtained by inefficient rectification of AC power, which again increases the capital investment required — all in all not a very cost-effective alternative to throttling.

The development of the thyristor — the solid-state equivalent of the thyratron thermionic valve — has made speed regulation of AC motors of all sizes economically feasible. The 'squirrel cage' induction motor, the armature of which is constructed of bars rather than mere wires, and which looks a bit like a cage, has a very low armature impedance and consequently loses little power in wasteful heating. However, the torque/speed relationship is as shown in Fig. 9.15(a), from which it can be seen to be effectively a one-speed machine. By increasing the armature resistance a little as shown in Fig. 9.15(b) a near ideal torque/speed relationship can be obtained, and, although the additional armature resistance implies some increase in wasted energy (lower motor efficiency) and a slight derating to allow dissipation of the additional waste heat (a slightly larger motor may be required for a

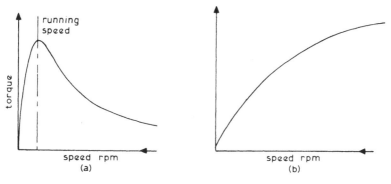

**Fig. 9.15.**

particular duty), these are small penalties compared to the enormous waste of energy implied by the inclusion of throttling devices in the process/plant system.

The 'final control element' in a variable speed drive system is termed a 'power controller', and consists of a trigger circuit, which is set by the process controller output, and a pair of thyristors per phase in the AC supply to the motor itself (Fig. 9.16).

There are three ways in which the power entering the motor is modulated by the power controller. The first is phase-angle control under which the thyristors are triggered to 'fire' (i.e. conduct) during only part of each cycle of the alternating current supply. That part can be increased and decreased so that the *mean effective power* input over a number of cycles is modulated in proportion with the control signal from the process controller (Fig. 9.17).

The use of this method of power modulation produces considerable

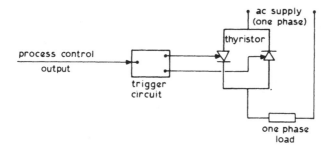

**Fig. 9.16.**

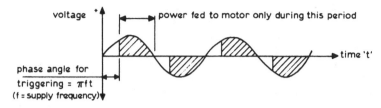

**Fig. 9.17.**

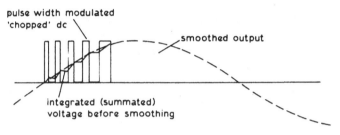

**Fig. 9.18.**

'harmonics' in the AC system owing to the irregular 'chopped up' nature of the output from the power controller thyristors, and these harmonics may cause signal interference. For large equipment, such as very large compressors, the expense of more complex solid-state power controllers may be justified: these convert the AC power supply to DC, smooth it using capacitors, and then 'chop' it up using pulse-width modulating inverters to provide an AC supply to the motor at variable frequency (Fig. 9.18). Such controllers are very expensive and complex, and contain control loops of their own, but they can drive standard squirrel cage motors at speeds which relate to the alternating frequency output of the controller (whereas with the simpler type of controller the motor speed is not related to the frequency of the supply, which is constant).

A simple form of power controller is sometimes used in which the supply is switched on and off, not during each cycle, but every few cycles (Fig. 9.19). The on/off ratio is varied within a fixed total period so that at one extreme no power at all is fed to the motor, whilst at the other the normal power supply is continuously connected. In between these two extremes the motor, of course, receives 'bursts' of power. This is a form of proportional time control and it is only suitable for systems in which there is a large inertia or flywheel effect to smooth the irregularities in power supply.

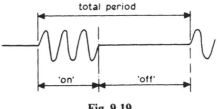

**Fig. 9.19.**

The design of motor drive controls is beyond the scope of this book; however, it has been mentioned here as in many cases there is no longer any excuse for the use of energy-wasteful throttling valves as final control elements. Pumps, fans, compressors and other motor-driven equipment can be controlled at least as well, and often better, by regulation of the drive motor speed.

# CHAPTER 10

# *Process Control and System Design*

## 10.1 INTRODUCTION

Processes are developed in the laboratory in the first place and later in small-scale pilot plant. After this stage full-scale plant is designed to carry out the process, not by instrumentation and control engineers, but by chemical process engineers. Automatic control systems are designed to assist with the operation of the plant in much the same way as human operators would do: often their function is merely to maintain a given steady state which is different from the natural steady state of the plant (level control on a vessel, for instance). Nevertheless, the way such regulating controls will react to disturbances, and indeed react with each other, has to be considered at the design stage. This behaviour will depend as much upon the design of the plant and the plant equipment items as on the design of the control system itself. The technician must learn how to diagnose unsatisfactory behaviour which will often be wrongly blamed on 'the instruments', as well as to diagnose faults in the instrumentation itself. The designer must learn how decisions taken by other design engineers or equipment manufacturers can seriously influence his ability to design a good control system, and must then apply this knowledge *early in the design process*, to ensure that the best compromises (all design is compromise) are made from his point of view. All this must come largely from experience; in this chapter it is only possible to introduce the concepts which will guide a well-educated engineer or technician who has wide knowledge not only of instrumentation and control principles, but also of the functions of other designers. In short, the good instrument technician or control engineer must be a jack-of-all-trades as well as master of his own.

This chapter will start by outlining the purposes of the various engineering drawings which the technician, as well as the engineer, must know how to use. It will go on to show how lack of consideration at the design stage can lead to operating problems, and later to show how system design should be approached. The later parts of the chapter will describe the design of typical systems.

## 10.2 PROCESS FLOW DIAGRAMS

The process flow diagram is a drawing which is intended to show the quantities and flow rates of material and energy in the plant. Usually values are shown at three different operating rates: minimum, design and maximum. The information is needed in the first place by design engineers, to enable the ranges of instruments, such as flow meters, pressure sensors and temperature sensors, to be fixed and to 'size' final control elements. This drawing will also show where flows start from and go to in the plant, and where control loops are required or measurements have to be made in order that the operator can do his job. Perhaps most important of all it will show the relative positions of plant equipment in piping systems, e.g. whether the control valve is upstream of the flow measuring device or downstream.

The process flow diagram will *not* normally show elevations, i.e. the heights above a datum (usually the ground) at which items of equipment are mounted. For this information the structural drawings will be required. Such information is often of vital importance in establishing the cause of faults, especially soon after start-up of new plant, because the design engineer may have overlooked some change in piping made late in the design. This may affect the working of a level, pressure or temperature measurement system (see Chapters 2, 3 and 6 for a fuller discussion), or it may mean that a control valve is now too large or too small.

The portion of a typical process flow diagram, shown in Fig. 10.1 (without process data), shows that the instruments and controls required are given only in enough detail to illustrate how the control systems operate. In the flow sheet shown in Fig. 10.2, for the same plant, a more complicated control system is shown; this has a level/flow cascade control for the receiver, and a cascade system with temperature controller as master and a flow ratio control system as slave to control temperature in the column itself.

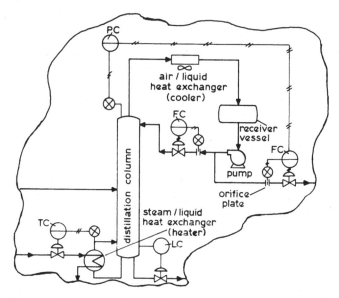

**Fig. 10.1.** Key: LC, level controller; TC, temperature controller; PC, pressure controller; FC, flow controller; O, local controller; ⊖, panel-mounted controller; ⊗, transducer; ——, process pipe; ╫╫, pneumatic signal connection; ——, impulse connection.

## 10.3 INSTRUMENT DIAGRAMS

The way in which the process is to be controlled is shown on the process flow diagram, but not the way in which the control system will achieve this. The instrument engineer or technician needs to know exactly what connections must be made between the different component items which make up the control system(s), where each item is located, where power supplies are to be connected, and details such as earth or cable screen connections, which do not concern the process operation. The existence of such items as valve positioners or booster relays, which are important for the operation of the control system, will not always be shown on a process flow diagram.

Once the process flow diagram has defined the way the process is to be controlled, a drawing called the 'instrument diagram' is produced to show how the instruments and controls will do this. There are a number of ways in which this can be done, the most usual being 'loop schematics'. It can be seen from the flow diagram in Fig. 10.1, that the

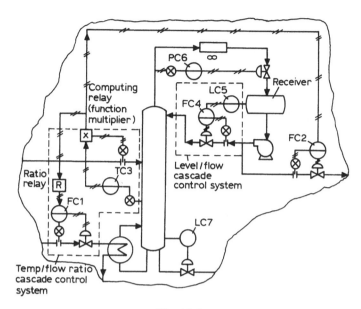

**Fig. 10.2.**

control system for the process unit (distillation column) really comprises four separate and independent control systems or loops. This is not an uncommon situation, but in contrast the flow diagram in Fig. 10.2 includes one system, the temperature/flow ratio system, which contains three controllers (TC, FC2 and FC1), two computing relays (X and R), three transducers and two control valves. A typical 'loop schematic' of this is shown in Fig. 10.3 for a pneumatic system: an electronic system loop schematic (Fig. 10.4) would not be very different, but might be more complex if intrinsic safety barriers have to be used to overcome the explosion and fire hazard.

It should be noted that in both the loop schematic (Fig. 10.3) and the process flow diagram (Fig. 10.2) (and in the specification) each item of equipment is given an identification number which identifies it with the measurement and the loop.

Figures 10.3 and 10.4 are divided into three areas: plant, back of panel and front of panel. The panel is usually built by a specialist contractor, and the loop schematics are usually all the contractor needs, together with a front of panel layout (to show where the indicating and recording instruments should be mounted) and the equipment specifications. For

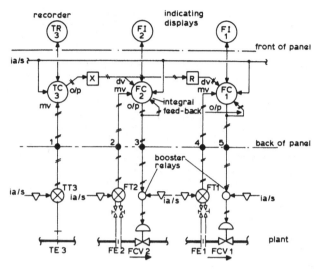

**Fig. 10.3.** Pneumatic loop schematic. Key: mv, measured value; dv, desired value; ia/s, instrument air supply; o/p, output (control action); FCV, control valve (flow); TT, temperature transducer; TE, temperature sensor element; TC, temperature controller.

this reason, connecting terminals for connections to the plant are shown and numbered. The construction engineer will use the loop schematic, together with the equipment specifications and detailed engineering drawings, to show how each item should be mounted or installed in the plant.

Once the plant is built and commissioned the process flow diagrams and loop schematics should be corrected to take account of any changes made during construction and commissioning, as should the equipment specifications (the 'trim' of a control valve may have been changed, for instance, because it was found, during commissioning, to be too large). All these documents become the record of the plant and are used by the maintenance staff; if they are not corrected and then kept up to date it will be difficult to know what is actually installed, and therefore to maintain the plant properly.

## 10.4 OTHER DRAWING RECORDS

In addition to the process flow diagrams and loop schematics (or other form of instrument *diagrams*) several other types of drawing have to be

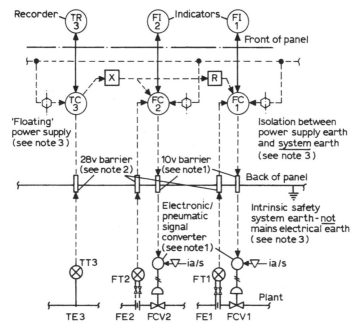

**Fig. 10.4.** Equivalent electronic loop schematic. (Integral feedback not shown — would normally be an internal connection.)

*Note 1:* Electronic/pneumatic converters are devices which operate on the force-balance beam principle to produce a pneumatic analog of the electronic (analog) signal. The type of barrier used will depend upon the impedance of the force-balance coil in this device.

*Note 2:* The dotted lines represent *two* core connections: one core is in each case connected to *system* earth (not the same thing as power supply earth). Thus, transducers, electronic/pneumatic signal converters and controller input and output circuits are all connected to system earth through the intrinsic safety barriers.

*Note 3:* The floating power supplies isolate the instrument system from the power supply earth. This satisfies the requirements of intrinsic safety and also prevents electrical 'noise' from entering the analog signals through 'earth return loops'. Care should be exercised to ensure that the analog circuits are not connected to the control panel, and hence to power supply earth (the panel will almost certainly be earthed in this way).

produced at the design stage. Of these the signal piping interconnection drawing or schedule informs the construction and commissioning engineers of the route to be followed by each pipe or cable from the central control panel to the plant-mounted items (transducers, control valves, etc.). Each connection may pass through several junction boxes and be made through one core of a multicore cable or pipe over part of the distance. Drawings are also needed to show how sensors are to be mounted in vessels or pipes, etc. (to ensure that the temperature sensor is inserted into the pipe the correct depth, for instance). These drawings too, if kept up to date, are of great help to maintenance staff and others after start up of the plant.

Finally, at the design stage the installation of such items as sensors, control valves and impulse connections have to be detailed on the piping drawings as they become part of the plant itself. *The instrumentation engineer should always ensure that installation is such that the equipment will operate as intended.* Ensuring that the pipework upstream of a flow-measuring sensor will provide suitable flow conditions is a case in point, i.e. ensuring that control valves are installed *downstream* of flow measuring sensors, or that sensors and final control elements are not so far apart that appreciable dead time is introduced into the control loop because of the time taken for process material to travel between the two (see below). All these drawings should be available, and may help maintenance staff to find the cause of faults which are not simply caused by failure of instrument equipment.

(Note that in the system shown in Fig. 10.4 integral control action has to be used on the two flow controllers even though they are 'slaves' to the master temperature controller. This is because they are also part of a ratio control system and therefore the flow rates in both cases must be controlled at the desired value (i.e. there must be no offset). Feedback of the controller output signal may have to be connected externally so that some form of integral action limiting can be applied. Details of the actual controllers and other items must be obtained from the manufacturers before the loop schematics can be completed.)

## 10.5 PLANT DESIGN FOR CONTROLLABILITY — STEADY STATE

As was pointed out in an earlier chapter, the plant itself is part of any control system. It has already been shown (in Chapter 9) that the 'installed' characteristic of a final control element depends on the plant performance characteristic as well as on the performance characteristic of the valve itself. The purpose of a control system is to 'manipulate' some variable of the process and it must not be overlooked that there will be finite limits beyond which this manipulation will be *impossible*.

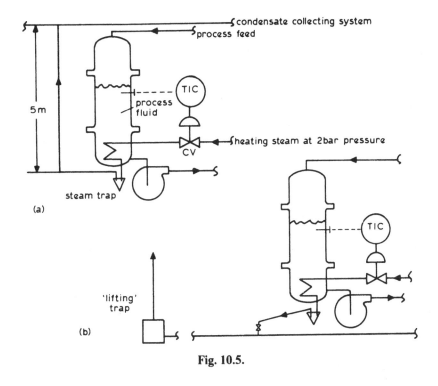

**Fig. 10.5.**

As a simple and not unusual example, consider the diagram of a heating coil in a vessel shown in Fig. 10.5.

The rate of heat input to the process fluid depends upon the temperature difference between the steam condensed in the coil and the process fluid. Steam at 2 bar (gauge) pressure is at a temperature of 134°C. Regardless of the size of the control valve no more steam can enter the coil than can be condensed at a temperature difference = temperature of process fluid −134°C.

At the other end of the range, no matter how wide open the control valve may be, no steam at all can condense (in steady state) unless the condensed steam can be discharged. In the system in Fig. 10.5(a) condensed steam will cease to be discharged when the pressure in the coil is less than 0·5 bar, at which pressure steam temperature is 112°C. Thus, if the temperature of the process fluid is 100°C the range of the heat input and therefore the process feed flow rate, can be no more than:

$$\frac{134 - 100}{112 - 100} = 2 \cdot 83 : 1$$

and this has absolutely nothing to do with the sizing of the control valve. In fact, the heating coil will have been oversized by the utilities engineer for duty at design process flow rate, to allow a margin of safety in his calculations and for fouling (reduction of heat transfer owing to deposits on the steam coil surface — inside or outside). Thus, it is quite possible that the range of operation will be considerably less.

When, during operation, the process feed flow rate is reduced below the minimum rate thus established, the temperature control will start to behave in a most unsatisfactory manner. Once control action has caused the control valve to close to the position at which the differential temperature is 12°C and the pressure in the coil 0·5 bar (gauge), condensate will cease to flow out through the steam trap and will accumulate in the coil. After a time (depending on the volumes in the coil and in the process vessel) this condensate will cool to 100°C and the *effective* heat transfer surface of the coil will be reduced to the extent that it is full of condensate at the same temperature as the process fluid. Eventually after further delay, the whole coil would be full of condensate, and no heat would be transferred to the process fluid. However, before this extreme position can be reached, the temperature of the process fluid will have fallen and the controller will have reacted by opening the control valve, with the result that pressure will have risen in the coil and *all* the condensate will have been expelled. Depending on the rate of steam condensation required to maintain the temperature of the process fluid at the desired value (which will in turn depend upon the temperature and flow rate of the process feed) and the volume of the steam coil, the controlled temperature will be seen to cycle in such a way that it will probably be said that the temperature control is not working properly. It should be realised, however, that none of the things that are happening have anything at all to do with any part of the instrumentation: it is the design of the heater system which is at fault, but it is the instrument technician who will be expected to find out what is wrong.

The second system, shown in Fig. 10.5(b), is much less likely to give the same trouble. Condensate is discharged to a low-pressure collecting system, without having to overcome the static head in the first system (Fig. 10.5(a)). Thus, the pressure in the coil can fall to almost atmospheric before the cyclical behaviour described above will set in; the range of process feed rates is then:

$$\frac{134 - 100}{101 - 100} = 34{:}1$$

or more than 10 times as great. If the condensate system cannot be altered to operate in this way, then steam at a higher pressure must be used (which would mean that the steam coil would have to be 'rated' for use at the higher pressure). If 10 bar steam pressure were used, for instance, the range would be:

$$\frac{184 - 100}{112 - 100} = 7{:}1$$

which is nearly three times the range possible with steam at 2 bar pressure.

Of course, had the desired temperature of the process feed been higher, say 110°C instead of 100°C, the problem of cyclical behaviour might never have been experienced, since then the range of feed flow rates would be:

$$\frac{134 - 110}{112 - 110} = 12{:}1$$

which would probably be quite adequate. Thus, such a problem may arise, not because of bad design but because a process operator has decided to change the process operating parameters.

The reader should learn from this simple example first to ask himself, when looking for the reason for bad control, is it possible for the control system to do what is expected.

## 10.6 PLANT DESIGN FOR CONTROLLABILITY — DYNAMIC RESPONSE

Even if the sort of problem described above (which makes it impossible to establish steady-state, or equilibrium, conditions) does not occur, there are many ways in which control can be adversely affected by design or operation of the plant, without there being anything wrong with the instrumentation. This is because of the delays which occur in the process and measurement, as was shown in Chapter 8. Consider the system shown in Fig. 10.6, for instance.

Heat transfer between the heating fluid and the process fluid, and also from the process fluid to the temperature sensor, will make the response of the process and measurement second or third order. However, in addition, further dead time will be added to this response (decreasing $T_c/T_d$ in Fig. 10.7 and hence making control more difficult)

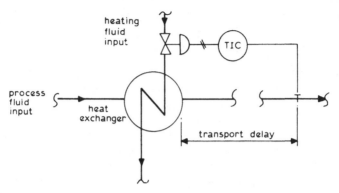

**Fig. 10.6.**

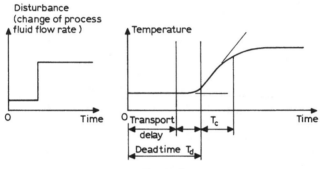

**Fig. 10.7.**

by the 'transport delay' introduced by the distance the temperature sensor has been located downstream of the heat exchanger (Fig. 10.7). This distance may be large because the local indicating controller has been sited in a suitable position to be seen by operating staff, whilst the control valve and heat exchanger are in an elevated and relatively inaccessible location. The sensor controller distance is limited by the available impulse connection (which is fixed when the instrument is ordered — perhaps before design is completed) but the controller-control valve distance is not constrained in this way. Hence, all these factors may result in a considerable length of pipe between the heat exchanger and the temperature sensor. This is doubly unfortunate because not only is control made more difficult under any conditions by the unnecessary addition of dead time, but this dead time itself will vary with the flow rate of the process fluid, which is probably a function of the plant throughput.

## 10.7 DYNAMIC NON-LINEARITY

The previous section showed how control can be made much more difficult by the unnecessary addition of dead time to the system response. It should be clear by now that the design of the plant equipment items is just as important for good control as the design of the controllers or final control elements. Dead time and lumped time constant which characterise the response are largely determined by 'residence time', i.e. the time that on average a small unit of flow 'resides' in a unit of plant. For instance, in the example given above, at any particular process fluid flow rate a small quantity will remain in the body of the heat exchanger for a given time (the residence time) depending on the capacity of the heat exchanger body and the process flow rate. If heat is being transferred from the heating fluid to the process fluid at a constant rate (depending on the temperature difference and heat transfer surface area and condition) then the rise in temperature of the process fluid passing through the heat exchanger obviously depends upon the residence time (or capacity) of the heater. Thus, it can be seen that the size (and other aspects of the design) of the heater will determine the response of the system to control action. It can also be seen that, since residence time changes when the process flow rate changes, the response will also change, just as the dead time in the previous section changed.

It is very important to realise that the response of a system will change with throughput, and therefore control action must be retuned for optimum results whenever the throughput changes substantially. This phenomenon is known as 'non-linearity', a linear system being one in which the response is constant over the whole range of throughputs — a comparative rarity.

The reader will recall from Chapter 8 that the integral and derivative control action factors are dependent upon the basic proportional action factor. Hence, if this is changed, the integral and derivative components of the control action change in proportion with the proportional action component. If, in the example in the previous section, throughput changes, the dead time and lumped time constant will both change approximately in proportion, and the response curve will change only

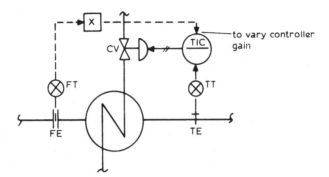

**Fig. 10.8.**

in that it will be 'stretched out' along the time axis (i.e. it will not change in form). Under these conditions it is only necessary to vary the proportional band (gain or proportional action factor) of the controller to retune it for the new throughput. Controllers are available which have this facility, and a suitable non-linear control system is shown in Fig. 10.8 that can replace that shown in Fig. 10.6 under conditions where the throughput is expected to vary over a wide range.

In this system a computing relay, X, could be adjusted by trial and error so that as the throughput changes, the gain of the temperature controller is increased or decreased by an amount which alters the controller response just sufficiently for the loop response to remain approximately constant. There are in fact microprocessors on the market today which provide such a facility.

## 10.8 POOR CONTROL AND BAD MEASUREMENT

Consider, for instance, the case of flow measurement in a pipeline which does not necessarily run full (Fig. 10.9). All sensing elements must be located in such a position that they are fully immersed at all

**Fig. 10.9.**

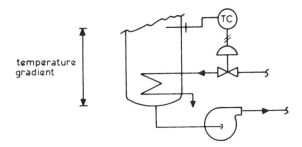

**Fig. 10.10.**

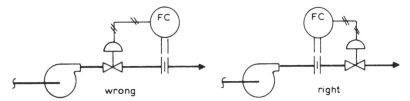

**Fig. 10.11.**

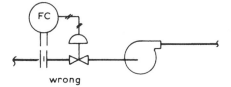

**Fig. 10.12.**

times and in all circumstances (including shut-down). Temperature-measuring elements may not be properly immersed to the correct depth, and pressure impulse connections can be blocked. It is no use measuring the process variable in the wrong place and then expecting the control system to work properly (Fig. 10.10). Nor should pressure tappings for level measurement in tanks be made too close to the bottom of the tank if sludge or sediment is likely to gather there.

The position of the control valve in the piping system is of great importance in most cases. It must never be placed upstream of a flow measuring element (see Fig. 10.11). It should not normally be placed in the suction of a liquid pump (Fig. 10.12). This is because the pressure is always lower in the suction. Two problems may arise from this fact: first as the control valve closes it may reduce the pressure downstream, that is at the pump suction, to such an extent that flashing occurs in the pump, spoiling its performance and thereby interfering with control functions (and quite probably damaging the pump also). Secondly flashing may occur in the control valve itself, making its sizing and characteristics indeterminate. A net positive suction head is always quoted for a pump, i.e. the minimum head pressure at the suction which will ensure that flashing does not take place. However, it is better, wherever possible, to locate the control valve in the higher pressure section of the pipework.

There are many ways in which measurement may be the cause of poor control; the instrument technician will have to learn by experience how to recognise them, and what to look for.

## 10.9 OBJECTIVES OF PROCESS CONTROL

Whilst the measurement and regulation of process variables is essential to control of the plant, there is a great deal more to the design of a control system which will make it possible for a modern process unit to be operated efficiently. In many cases operation without such a control system is not possible: modern processing units are often so complex that operation on the basis of human manipulation of all the process variable set points is out of the question. On the other hand, it must not be thought that the point has yet been reached where human influence is not required to operate the process units. In any modern plant there is a 'hierarchy' of control, with the human operator taking the most complex decisions whilst mechanisms take the simpler ones, such as when and how much to move the control valve stem in order to return the process variable to its desired value. In many cases, as has already been seen, control mechanisms decide how the desired value of a lower level controller should be altered to achieve a higher level desired value (master/slave cascade control). Increasingly today, digital computers are programmed to make decisions at a higher level than this; such decisions often involve changing the set points of several controllers in order to maintain some optimal processing condition, such as energy balance or product purity. Even without programmable computers, control systems have to be designed, which may incorporate many regulating 'loops', to carry out a given control strategy.

Whilst the design of measurement systems and the systems by which control is to be achieved are the responsibility of the instrumentation and control engineer alone, the design of the process control is not. Such design is not concerned with *how* control is to be achieved, but with *what* is to be achieved by the control system. Objectives must first be defined, process limits and constraints identified; it must be determined which process variables require to be and can be measured, and which can be manipulated. The measured and manipulated variables must be paired, the necessary feedforward/feedback, cascade, ratio or interactive regulating strategies identified, and the degrees of freedom of the process considered. Such design must be the joint responsibility of the process engineer and the instrument/control engineer; the former knowing best what is required, the latter what is possible.

Too often the control system for a process or plant is designed without due thought to the emphasis which should be given to, often conflicting, objectives. For instance, in times of recession it is often the case that a plant is operated below its maximum capacity, because the market demand is less than it was when the plant was designed. In such circumstances, the quantity which can be sold will depend upon whether the quality and price are better than or worse than those of competitors. Therefore, the objectives will be to operate the plant in such a way as to maximise productivity (i.e. the quantity of product made for given inputs of raw material (feed) and energy) and/or product quality, so that the best product possible can be sold at the lowest price possible. On the other hand, in a boom situation, it may be possible to sell as much as can be produced, even though the quality is not as high as it could be. Whilst judgements as to which objectives are appropriate are the province of management, the design of the control system must reflect these objectives, and often must be such as to provide as much flexibility as possible to cope with different sets of objectives.

Consider the block diagram of a chemical extraction process given in Fig. 10.13. The raw material from a mining process is digested (that is dissolved) in a liquid in order that other material (which does not

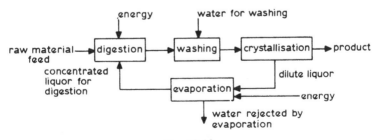

**Fig. 10.13.**

dissolve in the liquid) can be removed. The product is then separated out from the 'liquor' by crystallisation, and the remaining liquid is recycled to digest more raw material. Energy is necessary to heat the digestion process and to concentrate the liquor left after crystallisation. If the objectives of control/operation are to satisfy a market where less can be sold than can be made (i.e. recession), then it is important that as much raw material is digested as possible for each unit quantity of liquor. Roughly the same amount of energy will be used to concentrate it in the evaporation stage no matter how little or how much raw material is digested, and thus the cost per unit of product will be minimised. If on the other hand it is possible to sell however much is made, then it will be less important how much each unit costs, and indeed the additional profit from higher sales is likely to outweigh the slight reduction in profit per unit produced. In the first case the control system and operating techniques must aim to provide the maximum efficiency of digestion at the expense of throughput, whilst in the second case it is throughput which must be maximised even at the expense of efficient digestion.

The company making this product may find that it can sell it at a much higher price per unit if it is very pure. In this case control and operational objectives will be to maximise product purity even at the expense of both throughput and digestion efficiency, and possibly regardless of the economic climate. In practice, of course, the emphasis on the various objectives will change from time to time and will in any

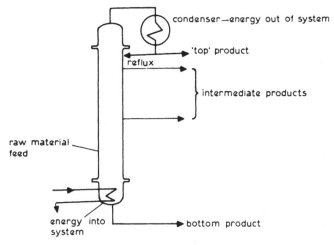

**Fig. 10.14.**

case be a combination of all these factors. For instance, it may be possible to sell only some of the product to the buyers of high purity material.

Distillation is one of the basic processes of petroleum refining (Fig. 10.14). Crude oil is separated into products which contain a high proportion of 'light ends' (the more volatile components) and those which contain a high proportion of 'heavy ends'. In general the top products are sold as petroleum and the intermediate as diesel or light fuel oil, whilst the bottom product is only suitable as fuel for industrial use. The market requirements and the 'quality' of the feedstock both change from time to time, and the control systems and operating techniques must be designed so that the distillation plant can produce more of one type of product and less of another as required.

To illustrate how differing objectives may affect the design of control systems, as well as plant, compare the processing requirements of a simple distillation process designed to produce 'top' product of maximum purity ('bottoms' probably being considered as waste product) and one designed to produce two products (top and bottom) of roughly equal values. Maximum recovery (i.e. maximum production of distilled products) is incompatible with minimal energy use. If one product is valuable whilst the other is essentially waste, the lowest cost per unit of product is usually obtained by using as much energy as possible in order to obtain the maximum separation, since the energy cost per unit of productivity is likely to be less than the cost of losing valuable product not separated from the waste stream. On the other hand, if the value of the two product streams is roughly similar, the lowest cost per unit of production will be obtained by minimising the energy used, since energy costs are a very significant part of the total cost of production these days; the fact that a significant component of each product stream is the unwanted product will not in many cases affect the product sale price. There are cases, of course, where purity of product composition is important, regardless of the value of each product, and in such cases the correct objectives may be more difficult to define.

## 10.10 OBSERVABILITY AND CONTROLLABILITY

The parameters of a process are measured in order to determine how the process is behaving and there must be sufficient information for this purpose. If there is not, the process is said not to be observable and therefore cannot normally be controlled. For instance, in order to control the process of steam raising in a boiler, it is necessary to measure the pressure of the steam and the flow rate of the steam leaving the boiler, so that the inputs of water and heat can be regulated. In practice, the rise or fall of the level in the steam drum is often taken as a measure of the mismatch between the quantity of water entering and the quantity of steam leaving. If the level is static, this satisfies the requirement for 'mass balance'. Pressure is a measure of the 'condition' of the steam; condition in this case indicating the quantity of heat added to each mass unit of water to convert it to steam. The rate of heat input is dictated by the condition and flow rate of steam generated.

If the mass balance control (mass flow of steam generated per unit time/mass flow of water into the boiler per unit time) is based on the measurement of drum level, this assumes that the mechanism for regulating the drum level operates perfectly, and as will be seen later this brings its own problems. Having considered the boiler plant itself, therefore, from the point of view of the requirements of control, it is necessary to consider the sub-units, such as the drum, to determine the needs of control. The selection, sizing and design of process plant (vessels, pumps, distillation columns, heat exchangers, etc.) are very much part of the design of any control system. The control engineer must therefore work closely with the process and other design engineers to establish control objectives, not only for the total plant but also for each individual piece of plant. These 'control objectives' are the critical requirements for good operation of the plant and the individual equipment items. Often, because of long delivery times, these requirements must be decided early in a design project, and the control engineer should advise the other design engineers of the constraints and requirements for good control as early as possible.

The basis of design of all process control systems must be the need to satisfy mass and energy balances in steady-state terms. In addition, recent increases in energy costs make it essential to regard minimisation of energy usage an objective of control. For instance, most of the energy wasted in distillation is due to incorrect control of the reboiler and condenser which, respectively, put energy into and remove heat from the process. The relationship governing the steady-state operation of a heat exchanger is given by:

$$Q = UA\,dT$$

where $Q$ is the heat transfer rate, $U$ the heat transfer coefficient, $A$ the heat transfer area, and $dT$ the log mean temperature difference between the process and the cooling fluids (Fig. 10.15). Control of the amount of

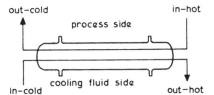

**Fig. 10.15.**

heat removed from the process can be effected, therefore, by manipulating any of the terms on the right-hand side of this equation, i.e. the heat transfer surface, the temperature of the cooling fluid (which amounts to manipulating the log mean temperature difference since the temperature of the process fluid is the controlled variable and cannot be manipulated) or the heat transfer coefficient (which is not normally feasible).

Water-cooled condensers normally used for this purpose are best controlled by 'tempering' the cooling water as shown in Fig. 10.16, using a recirculatory system in which the cooling water is circulated through the heat exchanger at a constant velocity: its temperature is regulated by manipulating the flow rate of the return, thus admitting more or less of the cold supply. The process fluid temperature is in turn regulated in cascade by manipulating the set point of the 'tempered' water. The

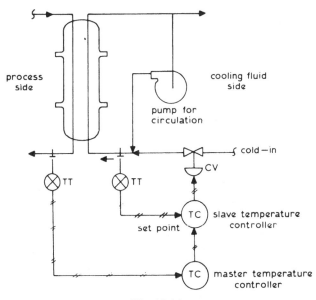

**Fig. 10.16.**

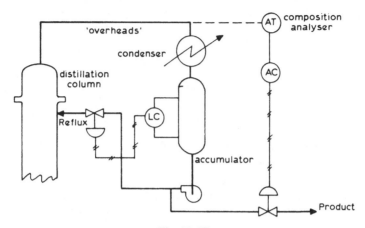

**Fig. 10.17.**

response of the 'slave' controller will of course depend upon the speed of circulation of the cooling fluid through the heat exchanger, which will in turn depend upon the sizes of the pipes, the pump capacity and the residence time of the fluid in the heat exchanger. Thus, the control engineer must work with the process engineer, the piping engineer and the mechanical engineer in specifying this equipment, bearing in mind that the response required of the master control must be several times slower than that made possible by these design features in the slave loop.

'Accumulator' vessels are often provided in the process design, as in the distillation column reflux system shown in Fig. 10.17, in order to give stability to operation. Unfortunately, they introduce many problems into the design of the control systems by reason of the phase lag introduced into the transient response of the system. Nevertheless, the same capacity is necessary to provide 'surge' capacity (temporary flow rates greatly in excess of the throughput capacity of the heat exchanger) to deal with process upsets. Hence, system design is bound to be a compromise and again it is essential that the control engineer is involved in the specification of the process equipment and vessels from the outset.

In the system shown in Fig. 10.17 the composition analyser is used as the basis of a control loop which regulates the top product flow rate to give a required purity of specification. If this analyser detects an increase in the 'impurities', the associated controller will act to reduce the forward flow, so that in steady state more of the condensed 'overheads' will be returned to the column as reflux, thus increasing purity. Unfortunately, this increase in reflux suffers delay due to the time constant of the accumulator vessel — the reflux flow rate will only increase when the level rises in the vessel owing to the reduction of flow rate in the top product line. This delay will depend upon the capacity of the accumulator, its shape (whether it is tall and slim or short and fat) and on the tuning of the level controller. This is typical of a system design which has been made by process engineers without the assistance of a control engineer, who would probably advise the use of feedforward control to overcome these difficulties as shown in Fig. 10.18.

In this system any change of flow rate in the product line (resulting from the action of the composition analyser control) will be 'fed forward' to reset the flow control loop on the reflux line immediately. Any change in the level in the accumulator vessel will in due course further modify the set point of the reflux flow controller in order to satisfy the steady-state mass balance. The equation of the computing relay will therefore be

reflux set point = level controller output signal − product flow rate

with appropriate scaling and biasing (to ensure that the level controller

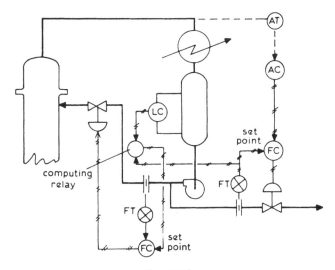

**Fig. 10.18.**

integral term is approximately mid-scale value when the system is operating in steady state).

## 10.11  DEGREES OF FREEDOM AND 'PAIRING' OF VARIABLES

A common mistake in the design of complex process control systems is to ignore the rule that the constraints on a system can never be more than the degrees of freedom (less, but never more). The degrees of freedom are the variables (measured or not) which can be manipulated in order to operate the process. These variables will in most cases be paired with the most appropriate measured variables — for instance, in the previous section the process fluid temperature (measured variable) was paired with the mean temperature difference in the cooling water circuit (the manipulated variable). To illustrate the point, it will be obvious that for a boiler generating saturated steam, only one or other of the steam pressure and steam temperature can be manipulated, and to attempt to manipulate both will be to break this rule. It is not always so obvious, however, that the rule has, or has not, been kept, and a distinction must be made between steady-state and transient degrees of freedom. For instance, in the previous section it was pointed out that the addition of an accumulator to the reflux system of the distillation process provided 'surge' capability: another way to look at this is that without the accumulator the sum of the product flow and the reflux flow must at all times equal the total overhead flow rate, whereas with an accumulator this constraint does not have to be observed transiently. The accumulator level controller is introduced to ensure that in steady state this constraint is observed. The feedforward action can then be tuned so that when the composition control loop causes a change in the forward product flow rate, the corresponding initial change in reflux flow rate is greater than that dictated by steady-state mass balance. Subsequently, the level in the accumulator will fall because of the imbalance, and the level controller will modify the reflux flow rate to avoid the accumulator running empty and eventually (by integral control action) restore the desired value, thus satisfying steady-state degrees of freedom. 'Squared-error' control action would allow small changes of level without significant control action, the control action factor increasing greatly, however, for large errors in level.

Once the degrees of freedom of the process system have been identified there will be a certain number of variables which must be measured (or analysed), and an equal number of (different) variables which must be manipulated in order to carry out the control strategy. Care must always be taken to ensure that they are 'paired' in such a way that each individual control loop manipulates the variable which most strongly affects the measured variable for that loop. If this is not the case, control will be poor if not totally ineffective, since control action taken in one loop will have a greater effect on some other measured variable than on that intended. Even if this rule is observed, it will sometimes happen that the effect of control action on other variables than that intended is very considerable; in such cases 'multivariable' control strategies are essential. Such interaction can be appreciated by considering the heat exchanger example in the previous section. If the flow rate of the process fluid is being manipulated in order to regulate some other measured value, such control action will inevitably influence the process temperature as will now be shown.

If the operating conditions are as shown in Fig. 10.19 with a process fluid flow rate of $F$, then the log mean temperature difference dT is

$$\frac{(100 - 60)}{\ln(100/60)} = 78 \cdot 3$$

and the rate of heat transfer is given by

$$Q = 78 \cdot 3 UA$$

If as a result of control action in the flow regulating loop the process flow rate is reduced 10%, then since $U$ and $A$ do not change

$$Q' = \frac{dT'}{dT} Q$$

and since $dT = 78 \cdot 3$

$$Q' = \frac{dT'}{78 \cdot 3} Q$$

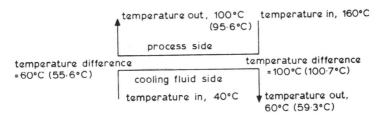

Fig. 10.19.

where $Q'$ is the new rate of heat transfer and $dT'$ the new log mean temperature difference. By iterative calculation the new values of inlet and outlet temperature on process and cooling fluid sides shown in brackets in Fig. 10.19 can be calculated. Thus

$$dT' = \frac{(100 \cdot 7 - 55 \cdot 6)}{\ln(100 \cdot 7/55 \cdot 6)} = 75 \cdot 8$$

It can be seen, therefore, that the process outlet temperature will change from 100°C to 95·6°C as a result of a 10% reduction in the process flow rate, unless the temperature control loop acts to correct this error. Thus, the action of one control loop has resulted in a disturbance arising in another control loop. This is known as interaction, and the ratio of the steady-state change in the process outlet temperature, which would result from a given change in the process flow rate if the temperature control loop were not operating, is referred to as the steady-state open loop process gain between the manipulated variable (process flow rate) and the measured variable (process temperature). In the course of the system design these steady-state process gains must be calculated, and the measured and manipulated variables which show the strongest 'coupling' selected as the measured and controlled variables, respectively, of the single input/single output (SISO) control loops.

In extreme cases the coupling between one manipulated variable and a measured variable may be of the same order as that between another manipulated variable and the same measured value; in such cases a decoupling (or multivariable) control strategy may have to be adopted. This strategy consists of cross-connecting the outputs of the SISO controller in such a way that it manipulates both the manipulated (or controlled) variables: the control action factors are different for each manipulated variable, however, and the combined control actions are designed to give 'non-interacting' control. In the example above, for instance, the process flow-rate controller output could be made to change the set point of the slave temperature controller in the opposite sense to the disturbance which is introduced by the change in flow rate, i.e. to raise the cooling water inlet temperature so that the steady-state drop in process temperature from 100°C to 95·6°C does not in fact take place. The control system would then be as shown in Fig. 10.20.

## 10.12 DYNAMIC INTERACTION

Steady-state interaction can be avoided by good control system design, and it may sometimes be possible to *avoid* dynamic or transient

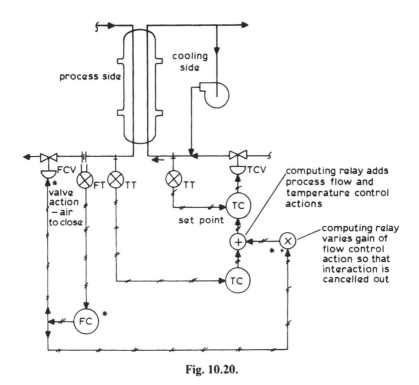

**Fig. 10.20.**

*Note:* The 'sense' of the flow controller output signal will depend upon the safety requirements (on pneumatic power supply failure) for valve operation. If the valve closes on increasing the signal (reducing the flow rate), then, since reduction of flow rate causes a fall in process temperature (interaction), the 'decoupling' control action is required to raise the cooling water inlet temperature in order to reduce the log mean temperature difference and thus counteract this interaction. Thus, the signal to the computing relay is of the correct sense in this case.

It should be borne in mind that the response of the flow control loop and that of the decoupling temperature control action will be different, the latter depending upon the rate of circulation of the cooling water through the heat exchanger, the former only on the speed of response of the control valve. It may therefore be expedient to limit the speed at which the flow control valve can respond to correspond with the speed of response of the temperature slave loop; in this way transient interaction between these two loops will also be avoided. It is clear from this that the transient response of the whole system is limited by the design of the heat exchanger.

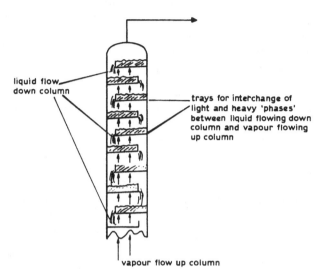

liquid flow
down column

trays for interchange of
light and heavy 'phases'
between liquid flowing down
column and vapour flowing
up column

vapour flow up column

**Fig. 10.21.**

interaction by ensuring that loop responses are approximately the same — by tuning. However, it is not always possible to achieve this, as the process itself may have more than one response to the same control action, and these responses may themselves interact.

For example, consider a change of reflux flow rate and its effect on the operation of the distillation process taking place in a distillation column having a number of trays (Fig. 10.21). Each 'tray' contains a small quantity of liquid in order to provide the residence time required for the interchange of light (more volatile) phases or components to the rising vapour and heavy (less volatile) phases to the falling liquid. Any change in the flow rate of the liquid flowing down, or of the vapour flowing up, will change the composition of the top product and the bottom product streams, but this change will be subject to the total delay represented by the many time-constant delays of the individual trays in series, and thus to considerable dead time (depending on the size of the column and the number of trays this can be of the order of hours). However, the hydraulic response of the system — i.e. the change in liquid flow down the column after a change in reflux flow rate — is very much quicker. This is because the flow of liquid from tray to tray is regulated by weirs on each tray, and there is little if any residence time delay (change in flow rate off the tray is almost instantaneous after a change of flow rate onto the tray). The increased liquid flow down the column will have to be matched by an increase in vapour flow rate up the column if the composition of the bottom product is not to be seriously upset; yet this control action will precede, perhaps by hours, the change in composition at the individual trays, which can progress only slowly in cascade down the column. Thus, whilst this disturbance is propagating slowly down the column, a new disturbance in composition will start to propagate upwards from the base of the column, and disturbed composition is inevitable throughout the column for some time. It is essential that the control system design recognises such interaction *within the process* and that it acts in such a way as not to exacerbate it. One way that this can be achieved is to use a sampled data technique: the measurement is 'sampled' and the appropriate control output produced by a conventional SISO controller, then no further change in this output is permitted until a time has elapsed which is rather longer than the system 'settling' time which includes dead time. A mechanism for achieving this is shown in the diagrams in Fig. 10.22.

Even in the absence of dead time in the system, system responses can often interact to provide a very difficult control problem; a prime example of this is to be found in the drum level control of the modern water tube boiler. A boiler is always demand-controlled, that is to say its function is to supply steam of a given quality when the process demands it. An increase in demand for steam requires a corresponding increase in the flow rate of feed water to replace the steam: however, in order to satisfy the steady-state mass balance, the flow of water is basically regulated by the drum level control loop. The interaction occurs because there are two separate and unfortunately contrary responses to the increase in heat input which must also occur when steam demand increases in order to satisfy the energy balance. The first response, which occurs very rapidly, results from the physical fact that the water expands in the boiler tubes when the heat input rate increases; the capacity of the tubes is much greater than that of the drum and so the level in the drum rises rapidly at first. The drum level controller, seeing the level rising, acts to reduce the feed flow rate initially; however, expansion of the water is a finite and therefore transient response to a step change in steam flow demand, and very soon the combination of increased outflow of steam and reduced inflow of feed water results in the drum level beginning to fall at an accelerating rate. This is the steady-state response to the increased demand, and unless the control

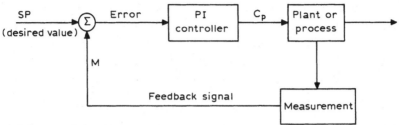

(a) Conventional feedback control

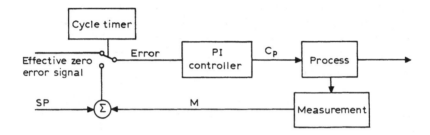

(b) Sampled-data feedback control

(i)

"On" state

(e = Error)

SP                                                            M

Hold state

(e = Error)

SP                                                            M

(ii)

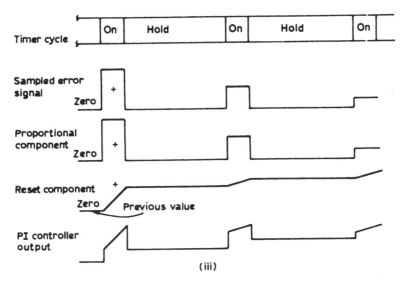

**Fig. 10.22.** (i) The feedback signal in conventional control (a) is compared in a summing junction with the set point; the resultant error, e, directly feeds the controller. In analog sampled-data control (b) the controller operates on a sample of the measured error signal during a brief period determined by the cycle timer, and then holds its existing output signal when the timer connects its input to an effective zero signal. (ii) When the desired set-point value, SP, and the measured variable, M, are briefly connected to the respective input ports leading to the summing point, the controller receives a sample of the actual error signal in an 'on' state. When switch action connects its inputs together, the controller receives an effective zero error signal and enters the 'hold' state. (iii) The automatic reset control function of the two-mode controller, third line down, displays a special function by holding the response to the actual error signal for the duration of the 'hold' period.

system is so designed that it is wary of the initial transient response, the level in the boiler drum will fall so rapidly that there will be no water in the drum before the feed water valve (which is often large) can open enough to check the fall. The solution to this problem is a combination of system design and adaption of the standard control actions (see Fig. 10.23).

Feedback regulation of the drum level is provided by a cascade system comprising a master level controller and a slave flow rate controller. Feedforward control is added by a flow rate controller on the steam flow, which, together with the flow rate controller on the water flow, constitutes a ratio control system. Steady-state dominance of the

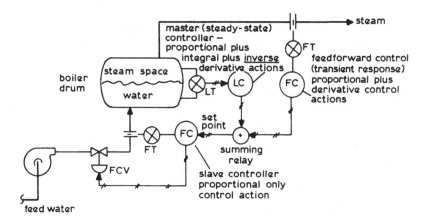

**Fig. 10.23.**

level control is ensured by the fact that the steam flow rate controller does not provide integral control action, whereas the level controller does. In order that the level controller will substantially ignore the initial rise in drum level, caused by expansion of the boiler water, *inverse* derivative control action is added to the proportional and integral actions: as its name implies, inverse derivative control action reduces the proportional action to an extent relative to the rate of change of the measured variable (level in this case). Thus the system ensures that when the steam flow rate increases, the water flow rate is immediately increased despite the rise in level due to expansion (the drum must be designed to accommodate this 'swell'), and that after this transient response has died away control of drum level dominates the water flow rate, ensuring that in steady-state the level will be at the desired value.

## 10.13 INSTABILITY

It has already been stated that process plant is normally designed to have as much self-regulation as possible; however, there is one group of processes which cannot be so designed — exothermic reaction processes. Such a process is open-loop unstable, that is to say if it is not controlled in some way it will 'run away' to some limiting state, perhaps even explode. This is because any rise in temperature causes the process to produce more heat and, hence, raise the temperature still further. The control system must increase the rate of heat removal when the measurement system detects a rise in temperature, but whether the closed-loop system will be stable or not will depend on whether the measurement system response is fast enough and whether the control action can be applied fast enough. If these responses are too slow the reaction temperature will increase faster than the control system can deal with it, and the closed-loop system will also be unstable. Such reactions take place in jacketed vessels, with cooling water circulating through the jacket, and response to control action (increasing the flow rate of the cooling water, or reducing the inlet temperature) is subject to the time-constant delays always associated with heat transfer through vessel walls, as well as delays due to mixing within the process materials inside the vessel. Temperature measurement, too, is subject to heat-transfer delays. Where there is a vapour space inside the reactor vessel and the reaction products are volatile (such as polymerisation reaction), a pressure sensor can detect increase in reaction rate more quickly than a temperature sensor, and a cascade loop as shown in Fig. 10.24 may provide a fast enough measurement response. However, the problem of response to control action is largely a matter of the design of the vessel jacket, cooling system and mixing equipment. For instance, the greater the temperature difference possible between the cooling fluid and the process material and the larger the effective heat-transfer surface, the greater the control action that can be applied and therefore the faster the

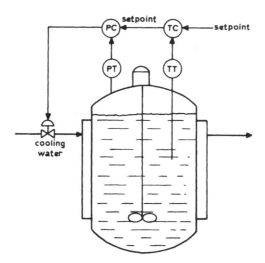

**Fig. 10.24.**

response possible. The capacity of the jacket will add a capacitive/
resistive delay to the heat-transfer process and should therefore be as
small as possible. However, equipment decisions of this sort are often
made early in the project design phase before a control engineer has
been able to evaluate the constraints.

The rate of reaction is proportional to the temperature at any time,
and writing this mathematically

$$WC_p \frac{d}{dt} = t$$

where $t$ is the temperature, $W$ the weight of process material, and $C_p$ is
the specific heat of the process material; or in operator form

$$WC_p pt = t$$

Replacing the operator p with the Laplace operator (to obtain a steady-
state as well as transient response) the transfer function is

$$\frac{1}{WC_p s} \qquad \text{or} \qquad \frac{1}{WC_p}\left(\frac{1}{s}\right)$$

In other words the *open-loop* response to a disturbance in $t$ is given by

$$\frac{1}{WC_p}\left(\frac{1}{s}\right)t$$

The mechanism of heat transfer will contain two capacitive/resistive
delays, one for transfer of heat from the process material in the reactor
vessel through the metal wall of the vessel into the cooling water, and the
second due to the capacity of the jacket delaying changes in the mean
temperature difference between cooling water and process material,
and therefore delaying the change in heat transfer rate. There will also
be some dead time due to transport of cooling water, mixing in the
process and other causes. The transfer function will be of the form

$$\frac{e^{-Ls}}{(sT_1 + 1)(sT_2 + 1)}$$

Similarly the transfer of heat through the wall of the thermowell into
the temperature sensor will add two more capacitive/resistive delays,
making the total transfer function for the process, heat transfer and
measurement

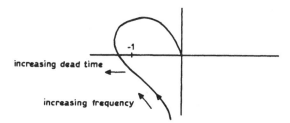

**Fig. 10.25.**

$$\frac{1}{WC_p} \cdot \frac{e^{-Ls}}{s(sT_1 + 1)(sT_2 + 1)(sT_3 + 1)(sT_4 + 1)}$$

The Nyquist diagram for this is of the form shown in Fig. 10.25 where it can be seen that encircling the −1 point can hardly be avoided. Note that 1/s in the transfer function adds a 90° phase lag at all frequencies. The importance of minimising dead time can be clearly seen from the Nyquist diagram.

## 10.14 RANGEABILITY

There is very little point in designing a control system to manipulate the right process variables in the right way, if in fact the variables cannot be varied over the full range required. Whilst this may seem obvious it is in fact too often overlooked in the design of control systems which incorporate more than one 'loop'. For example, consider the blending system shown in Fig. 10.26. The flow coming forward from a previous processing unit has a 'wild' composition, i.e. its composition varies in a random fashion with time. A side stream of different composition is blended with the wild flow to achieve the required specification. The flow rate in the 'wild' stream may vary as well as the composition, and the flow rate required in the side stream will be a function of the compositions in all three streams and the flow rate of the wild stream.

The ratio of the side stream flow rate to the wild flow rate is maintained by the ratio controller so that changes in flow rate in the wild stream do not disturb the composition of the product. The analyser controller resets the set-point value in the ratio controller (cascade) to maintain the product composition by feedback control. The ratio of

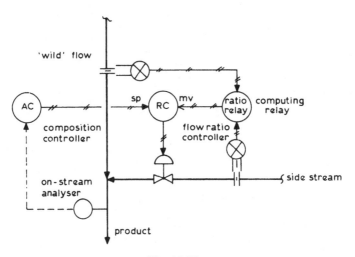

**Fig. 10.26.**

flows required is a function of the composition of all three streams and can be defined thus:

$$A_p(Q_w + Q_s) = A_w Q_w + A_s Q_s$$

where $A$ is the composition (in suitable units) and $Q$ is flow rate. The subscripts refer to the streams. Hence

$$A\left(1 + \frac{Q_s}{Q_w}\right) = A_w + A_s\left(\frac{Q_s}{Q_w}\right)$$

or

$$(A - A_s)\left(\frac{Q_s}{Q_w}\right) = (A_w - A)$$

or

$$\frac{Q_s}{Q_w} = \frac{(A_w - A)}{(A - A_s)}$$

Thus, it can be seen that if $Q_w$ (the flow rate in the wild stream) increases, $Q_s$ the manipulated variable must also increase. It can also be seen that if $A_w$ (the composition of the wild stream) is close to the desired composition $A$, the ratio of flows will be small; if very different the ratio will be large. Finally, it can be seen that if $A_s$ (the side stream composition) is close to the desired product composition, the ratio of side-stream flow rate to the wild-stream flow will be large. Thus, at one extreme the side-stream flow rate (the manipulable variable) will be very large if side-stream composition is close to product composition, wild-stream composition is very different from product and wild flow rate high. Alternatively side-stream flow rate will be very low if these factors are reversed. The rangeability required in the side stream must be carefully assessed early in the design because it is quite easy to find that it is not practicable, in which case it will be necessary to review the process design. It must be appreciated that this is not something to do with the control system design except that it may not be possible to design a suitable control system: other designers are unlikely to realise this and will probably blame the control system designer even though the decisions leading to such a state of affairs were not his.

# PART 3

# CHAPTER 11

# *The Development of Process Control Mechanisms*

The history of the development of process control mechanisms is synonymous with that of instrumentation in these industries. For this reason those who are most intimately concerned with process control are often the 'instrument' engineers. As the mechanisms available have developed, control of the process and plant as opposed to mere regulation of the measured parameters has developed, so that process control today is the province of the process engineer and even the manager rather than the instrument engineer. The developments in equipment systems over the last decade have been largely due to the advent of the microprocessor, and the pace has been such that the technical education of process control engineers is now woefully inadequate. The process control engineer, if indeed there is such an animal at all, must be a competent chemical, metallurgical, mechanical or electrical engineer first, a control systems engineer second, an electronics/communication engineer third, an instrument engineer fourth and also a competent manager.

In the latter years of the 19th century, process plant in chemical, metallurgical and other industries was simple. Instrumentation was, if anything, even more rudimentary, consisting of a few local measuring devices such as bourdon tube pressure gauges, liquid level sight glasses, liquid-in-glass thermometers, etc. Control mechanisms were largely confined to speed regulators on steam engines used to drive machinery. The human operator was the regulator, changing the position of throttling valves or dampers to achieve the desired value of important parameters of the process, or part of the process for which he was responsible. In short he was the process engineer and manager for a small plant such as the boiler. Communication with other process operators was often rudimentary, and the efficiency of process operation must have been very low indeed. As processes themselves became better understood and the operating requirements more complex, the demands on the operator became such that there was a considerable need for automatic regulation of the process parameters, so as to leave him to manage his plant rather than merely operate it. These automatic regulating mechanisms came about, in the first instance, by the addition of *position* servo mechanisms to local instruments. Indeed right up to the late 1950s such process controllers were still being installed in large numbers.

By the 1960s the size and complexity of many processes had grown to such an extent that *remote* servo regulation was required, so that the operator and the control mechanisms could be located in a large central control room which was often a long way from the plant being controlled. The operator now had very extensive responsibilities, and in some industries rarely set foot outside this control room. The *force-balance* servo control mechanisms themselves, whether they operated on electronic or pneumatic principles, had been developed to a high degree of sophistication though there was a problem with transmission time in some industries. The ergonomics of the very large number of indicator displays in such central control rooms became, by the late 1960s, almost ludicrous. Special multiple displays were developed by the major instrumentation equipment manufacturers to cope with this problem (in some cases the measured value signal pointer of an indicating controller was hidden as long as the servo mechanism was maintaining its desired value). At the same time computers and computer-based display systems were being developed to overcome these problems.

By this time control systems no longer consisted of single-loop regulators alone. Integration of the single loops into complex control schemes led to the development of a whole new body of theory and practice, concerned not with the functioning of the servo mechanisms alone, but with the much more complex functioning of the processes and the regulated systems. Such indeed was the complexity of process system design by this time, that it was by no means uncommon to find, after a plant had been commissioned, that the system design was far from satisfactory. The analogue systems could not be 'revamped' and, for continuously operating plant, had to be accepted. DDC (direct digital control) altered all that because the interconnections between separate control loops existed now only as software in the computer and could readily be changed at any time.

Digital computation made it practicable to construct multiloop control systems which previously incorporated complex, inflexible and inaccurate analog computing relays, and which were in consequence often almost unworkable. It also made it possible to restructure multiloop control systems *after* the plant was commissioned and even while it was running. This enabled the process engineer to try out several control system designs when that conceived at the design stage proved less than satisfactory. This in turn has shown the need for 'decoupling' the process interactions which are the cause of bad control system design. Distributed processing makes this relatively simple to implement, though to date equipment manufacturers appear to have missed this point! However the big weakness of the DDC system was the fact that all processing was dependent upon one processing unit, the CPU (central processing unit) of the central computer that replaced many independent analog controllers. Even if a second computer was provided as back up, the danger remained that the entire system might 'crash' and plant operation be jeopardised as a result. System availability was actually worse than that of analogue systems! Even worse in some respects was the constraint placed on commissioning of plant by the need to get the computer control system 'up and running' before any of the control systems could be commissioned.

Distributed processing has altered all this: control has returned to local control rooms, whilst centralised operator interfaces and all the advantages of multiloop control system building and historical recording, first introduced with DDC systems, have been retained. Most importantly, self diagnostic systems combined with suitable redundancy and back-up strategies make these equipment systems more, not less, available than analog systems. Unlike DDC systems therefore, there are few if any disadvantages to distributed process control systems relative to analog systems, whilst there are many great advantages. Nevertheless the original spur to their introduction was the considerable saving in cable cost that their use offers in the typical large process plant such as a refinery or steel works.

Microprocessors are now incorporated into transducers, controllers, analysers and final control elements. Data recording and operator interface displays have changed totally from those that have been common for more than a decade, though these changes first came about with the development of DDC equipment systems based on monolithic computers. DDC systems had grave disadvantages compared to earlier analog systems in terms of availability: only when microprocessors made possible distributed processing, did the advantages of digital control over analog become overwhelming. Not all these advantages have yet been exploited fully; some have not been exploited at all!

# CHAPTER 12

# *An Introduction to Distributed Microprocessor Systems*

## 12.1 EVOLUTION OF THE PROCESSOR

The purpose of any computer system is to process data in some way, and the heart of any computing system therefore is the 'processor'. Before the development of LSI and VLSI (very large scale integrated) electronic circuitry — 'chips' — the centre of any computer was the CPU (central processing unit), which carried out all the data processing for the entire system. This was accomplished by sharing the processing time between all the tasks that such a single processor (monolithic) system had to carry out. The CPU was constructed from discrete (transistor/resistor/capacitor) circuit elements as shown diagrammatically in Fig. 12.1.

The design of the CPU determined to a very large extent the design of the particular computer, and in turn of systems based on that computer. Speed of processing was very important because of the multiplicity of tasks which the CPU was expected to handle simultaneously. Execution of programs was complicated by the need to hop from one task to another. This was particularly true in the case of real-time systems, such as instrumentation systems, in which some tasks have to be 'serviced' at regular intervals and thus interrupt and slow up the execution of other tasks. The computer often had to control the 'fetching' and storing of the very data that it was called upon to process: even before the advent of LSI circuitry, data gathering or 'front end' systems were constructed to operate independently of the data processing to avoid this particular problem.

The advent of LSI and more recently VLSI circuitry has made it possible to 'package' large sections of the CPU, and more recently entire processors, as one mass-produced 'chip'. Where previously a CPU

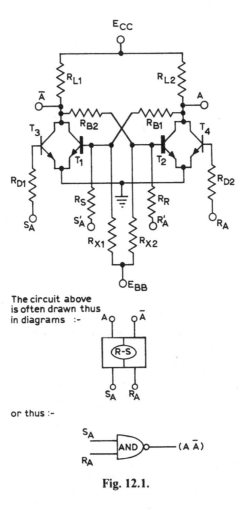

The circuit above
is often drawn thus
in diagrams :-

or thus :-

**Fig. 12.1.**

would have comprised a number of circuit boards crammed with logic elements and circuit components (resistors, transistors, etc.), a modern version would comprise one board filled with a number of large and small, but standard mass-produced chips. The modern microcomputer will employ a single 'processor on a chip', though it will be supported by other chips with standard functions in most cases. The cost of such computers is several orders less than that of older mini computers which they exceed in capability. It cannot be expected that they will rival in ability the larger 'mainframe' computers, but these are no longer used in such 'real-time' systems as instrumentation. The original DDC (direct digital control) systems which employed a single monolithic computer have been superseded by 'distributed systems'.

The change in instrumentation systems from centralised processing to distributed processing is perhaps as significant as that which has taken place in the electronics industry. The 'distributed' instrumentation system is a digital system, but unlike the older DDC systems it is not based on a single monolithic computer. Rather it is designed and built from a number of separate processing units each of which contains at least one microprocessor. Each such unit is a sub-system within the total system, but has its own separate functional specification. Communication between these separate units or sub-systems is obviously a very important part of the total system.

The cost of a microprocessor is such that one can usually be dedicated to a single task, so that it is idle to compare the speed of processing of a single microprocessor with that of a CPU in a monolithic computer. Naturally VLSI circuitry is employed in the construction of the CPU of a modern monolithic computer and the performance of such modern CPUs is very much better than earlier ones. The processing capabilities of a distributed instrumentation system cannot, however, be directly compared with those of an earlier DDC system using a monolithic computer. What can be said, however, is that by employing several or even many microprocessors within a computing system such as a distributed instrumentation system, in such a way that they share the many tasks of the system, the system performance can be far better than any monolithic computer-based system. Distribution of data processing also allows the system designer to distribute the functions of the instrumentation and control in a geographical sense also.

## 12.2 MICROPROCESSOR-BASED DISTRIBUTED SYSTEMS

A microprocessor-based distributed processing system, whatever its functional specification may be, can be defined as a computing system made up of two or more sub-systems each of which contains at least one independent data-processing unit (microprocessor). This definition effectively excludes any monolithic computing system, the functions of which are centralised by reason of the singularity of its data-processing unit (CPU).

The meaning of the term 'distributed processing' is clear; however it must not be confused with the term 'distributed instrumentation system'! In the latter not only is the data-processing function 'distributed' but so too are the instrumentation functions. The technology of distributed-data processing makes possible distributed instrumentation, the main advantage of which is that control functions can be located near the plant being controlled (rather than centralised in a single remote control room), whilst retaining the ability to interconnect control loops in ratio, cascade and other configurations. This greatly reduces the cost of cabling, typically 20% of the total in an analog instrumentation system, as well as improving the response of the system. At the same time measurement data can be transmitted to a central 'operator's interface' in digital form with no loss of accuracy due to transmission, and set point and control data can similarly be transmitted from the operator interface console to the local instrument systems.

It is important to appreciate that the modern 'distributed instrumentation system' has little in common with older 'DDC' systems, which replaced up to 400 SISO control loops with a single monolithic computer located in the central control room.

## 12.3 SUB-SYSTEMS OF A DISTRIBUTED CONTROL SYSTEM

The task-based sub-system in a distributed processing system will, at the very least, comprise:

    a processing unit;
    a clock;
    memory units;
    an application program (software stored in memory) bus, and bus
        buffers;
    an interface with other systems.

The sub-system must have at least one interface with outside systems and devices. The simplest interface is a 'serial port' but there are others which will be discussed later.

Few sub-systems will not include additional components, depending upon their particular functions. For instance, a data-gathering system may incorporate more than one processor; one for the task of 'scanning' multiple measured value inputs and placing their values in temporary memory locations, and another for the completely separate task of organising this data into 'packages' and transmitting them to other sub-systems over a 'shared' communication channel. Where more than one processor is incorporated into a single sub-system there will also have to be some 'buffer' memory added to hold data in transit between the processors, as their operations will have entirely different constraints (the reason for having more than one processor) and cannot be synchronised. A sub-system will often incorporate other chips whose functions are to enhance in some way the performance of the microprocessor/s: the industry produces such chips on a standardised basis wherever a sufficient market exists for them. Some of these chips will be discussed in detail later. Finally, yet other chips will be incorporated, the functions of which are of a specialised nature; multiplexing and A/D converters are required in a data-gathering sub-system for instance.

## 12.4 MEMORY

A number of different devices for storing program instructions and data are available. Data is changed often and it must therefore be easy to overwrite in memory. Whilst it is not usually too serious if data is accidentally overwritten, it is much more so if instructions are. This is because, whilst a program is readily rerun with the same data which is usually still available, to reconstruct the program will entail compilation and assembly. For these reasons data is 'written' into 'random access memory' devices (RAM); instructions however are sometimes 'written to' read only memory (ROM). RAM can be 'written to' and 'read from', but ROM can only be 'read from' as the name implies. (Note: addresses may form part of data or of instructions, depending on whether they are required to change or not.) ROM and RAM chips are available in capacities to suit most systems.

Other forms of memory device, such as soft and hard 'discs' and tapes, are not generally used in microprocessor systems as such, though they do usually form a part of computer systems, which are often attached to microprocessor instrumentation systems to complete a hierarchy of control.

## 12.5 THE TOTAL SYSTEM

The other components of a sub-system, clock, bus, interface and the specialised components used in distributed instrumentation and control systems, will be discussed in some detail in later chapters.

The total system will comprise many sub-systems. Some will be identical (such as data scanning and acquisition sub-systems), but there will be many different types of sub-system, each devoted to a separate task. The network of communication channels which connect the sub-systems together into a complete system are every bit as important as the sub-systems themselves, and the way in which such communication is achieved over common communication channels is fundamental to the performance specification and design of the system itself.

## 12.6 SYSTEM OPERATION

The processor carries out a series of 'instructions' which form a 'program'. Programs (software) are as much part of the equipment as hardware and are constructed so that the processor need only take each instruction in sequence and 'execute' it. In some cases the next instruction will tell the processor to 'jump' to some other instruction: in this way parts of the program can be automatically repeated (using different data) until some condition is satisfied when a 'compare' type instruction is carried out.

The process of 'fetching' an instruction into the processor and executing it has nothing to do with the instructions written by the programmer. As will be seen later, these activities are accomplished automatically by the processor itself carrying out a series of steps, each initiated by a pulse from a timing device called the system clock. At each such pulse the processor either connects itself to a location in memory at which data is to be found, or to which data is to be sent (addressing), 'fetches' data from or 'writes' data to such an address, or carries out some operation on data already stored within its (limited) memory. The process of addressing has been made possible by the development of the 'bus', which is a multiconductor connecting 'highway' rather like the domestic electrical 'ring main'. The 'bus' provides the means to connect the processor to other parts of the system, but it is necessary to have 'tri-state' switches in order that only one of these is actually connected up to the processor at any one time, and also that data can only pass in one direction at any one time. Hence it was development of the 'tri-state' electronic switch that made the microprocessor possible. During an 'address' time interval the processor selects and operates the required switches: during an 'execute' time interval it causes data to be transferred, in the required direction and to/from the selected device.

# CHAPTER 13

# *The Processor and Data Processing*

## 13.1 DATA PROCESSING

Data, or information, may be literal or numeric, i.e. letters/words or numbers. The computer only recognises two different states (voltage levels) usually notated '0' & '1', and so *all* data must be coded as 'binary' sequences of 0s & 1s. A letter requires 5 binary 'bits' to encode it, because the number of unique combinations of 0s and 1s is 2 raised to the power *n*, where *n* is the number of digits. ($2^5 = 32$, $2^4 = 16$, whilst there are 26 letters in the arabic alphabet.) The best known of such codes is ASCII which is widely used: to meet various operational requirements each letter actually uses 8 bits. Coding numbers is more straightforward, requiring a change of number base from 10 to 2. Difficulties occur with decimal points and round-off errors and a distinction has to be made between integers, such as house numbers, and decimal quantities. The processor must distinguish between literal and numeric data since the operations appropriate to each are entirely different. For instance:

$$9 + 6 = 15$$

whilst

$$\text{wolf} + \text{hound} = \text{wolfhound}$$

Further distinctions have to be drawn between data, addresses and instructions. Data, as already described can be literal or numerical; addresses comprise integer numerical data which is used for the special purpose of specifying the location of data in memory, so that the processor can locate the data or instructions when required to 'fetch' them or to 'store' them. Instructions are coded labels for processing operations which the processor is designed to carry out.

The unit of information (data) is then a 'bit' which corresponds to 1/8 of a letter, as already explained. Thus large amounts of binary data are required to represent quite small amounts of human information. Data is 'packaged' in 8 bits (a 'byte') or 16 bits (a word), so that a computer word corresponds approximately to a two-letter human word. Processors function by manipulating either a byte or a word simultaneously, which is referred to as parallel processing. Similarly bytes or words of data are transferred between component parts of the microprocessor system, in parallel, along multicore wires known as data highways or 'buses'. Each such 'data transfer' is regulated by 'timing pulses' from the system 'clock' and at each end of the bus, memory 'registers' of eight or sixteen bits must be provided to 'store' the data until it is transferred elsewhere. The process of data transfer takes place in two stages: first the appropriate memory register is 'addressed' during one 'clock interval', and then the data is transferred during the next. The process of addressing consists of 'setting' a group of eight or sixteen 'address lines', i.e. raising the voltage level to '1' on some lines whilst leaving it at the '0' level on others, thus creating an address word or byte which can be decoded and recognised by other components attached to the bus. This will be discussed in more detail in Section 13.4 and in Chapter 14.

## 13.2 THE PROCESSOR HARDWARE

The essential component parts of a processor are

1. The arithmetic and logic unit (ALU) which carries out the specified operations on data;
2. The logic control unit which interprets 'instructions' and sets control lines so that the necessary actions take place within the processor;
3. A number of temporary memory 'registers' which are used during the execution of an 'instruction' to store data.

The ALU and logic control unit comprise many transistor circuits in the form of VLSI chips. These circuits make up the 'logic' of these component parts of the processor. The logic has two functions, 'combinational' and 'sequential'. The combinational function of the logic circuits is to 'decode' address or instruction 'words' and thus enable the processor to set the required electronic switches. The sequential logical function is to make successive operations happen, as

| Inputs | | |
|---|---|---|
| A | B | Outputs |
| 0 | 0 | 4 |
| 0 | 1 | 2 |
| 1 | 0 | 3 |
| 1 | 1 | 1 |

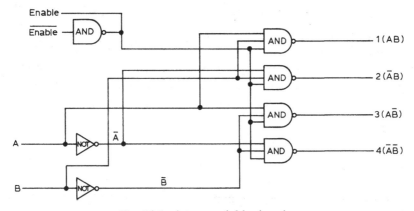

**Fig. 13.1.** A two-variable decoder.

decoded, in the required sequence. A very simple example is shown in Fig. 13.1. This shows, for instance, that switch number 1 will be 'set' if the value of two bits in an instruction word, A & B, are both '1'. If A is '0' (represented by the bar over A) then switch number 2 will be set instead. Only two circuits are used, an 'and' circuit and a 'not' circuit. The outputs obtained for each of the four possible combinations of '0' & '1' which bits A & B can adopt are shown in the 'truth table' in Fig. 13.1. The output changes from its last value, *not when the values of A & B change*, but when an 'enabling' pulse is received from the 'clock'. Thus the data set at the input to the logic circuits does not itself change the output from the circuit: this is done by the sequential function of the logic.

A logic circuit such as that shown in Fig. 13.1, but having eight inputs, can decode an eight-bit address or instruction word. As each successive word is presented to the logic, it produces a unique set of outputs to set the intended electronic switches on the address or control bus. The subject of design of such logic circuitry is outside the scope of this book: such design techniques are the basis for the design of all computer hardware.

## 13.3 PROCESSOR OPERATION AND SOFTWARE

Sequential execution of an instruction by the processor is evoked by a unique set of binary characters (e.g. 010011000) which the machine can recognise as code for the instruction. A 'program' may be prepared directly by the programmer writing instructions in this form: this is called 'machine language' and it is very rare for programming to be carried out at this 'level' today. Less rare is programming in 'assembler language' in which case each instruction is typed as a set of alphabetic characters known as 'mnemonic'. This enables the programmer to type his program on a conventional 'qwerty' keyboard. The mnemonics have to be translated into machine language by a process known as 'assembly' (see Section 13.5). However, most programs are written in 'high level' language in which a typed word, or set of words, describe an operation to be carried out on data. Each such operation may require execution of several assembly level instructions, hence high-level language programs must be 'compiled' into assembler code before being 'assembled' into machine code to be run by the processor.

Each single assembly level instruction is implemented in a sequence of micro-operations or 'steps'. Each such micro-operation in turn takes place in three stages, each stage being initiated by a 'clock' pulse. In the first stage the control unit sets up the control lines for the micro-operation; in the second stage data or address is 'set' onto the bus; and in the third and last stage the operation is 'executed'. The operation itself may be the transfer of data or address from a memory location outside the processor, or it may be an arithmetic or logical operation on data already in the processor, in which case there will be no need for the second stage and the processor may simply rest during this time. The 'operating cycle' of the processor will comprise a sufficient number of clock intervals to allow it to carry out any of these micro-operations in a 'clocked' sequence, as well as such 'housekeeping' functions as examining 'flags' or 'status' information or establishing a 'hold' (see Section 14.2). At the end of each processor cycle a new sequence of control signals will be initiated by the logic control unit to implement the next micro-operation. Each instruction will occupy a number of processor cycles, the more complex instructions taking as many as 20 cycles.

The number of clock intervals in the processor cycle and the design of the instructions constitutes the design specification of any particular processor and makes one processor different from another. The set of

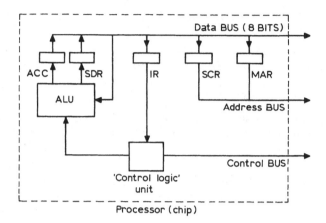

**Fig. 13.2.**

such instructions, typically as many as eighty, is termed the 'instruction set' of the processor. The logic unit contains all the circuitry necessary to both decode any instruction brought into the instruction register (IR), and to execute each of the micro-operations which are required, sequentially. Design of this circuitry determines how well the processor carries out its specified design functions.

Figure 13.2 shows the temporary memory registers of a typical processor, where

ACC = the accumulator — provides temporary storage for data;

SDR = store data register — provides temporary storage for data read into the processor from memory;

IR = instruction register — storage for the instruction word;

SCR = sequence control register — holds the next address for 'relative' address instructions (see Section 13.4);

MAR = memory address register — used to store the address for data, which is to be 'read', to be used in a subsequent operation, or to which the resulting data is to be 'written'. This register can also be used as temporary storage for data read in, as the SDR is, when both pieces of data to be operated on are obtained from memory. (In many cases one piece is the result of the previous operation and is already in the ALU).

The series of machine code instructions which constitute the program to be run by the processor are brought into the instruction register one at a time, decoded and executed.

Operation of the processor in carrying out the following assembly level instruction:

ADD A (x)

which calls for the data at (x) to be added to that already in the processor's accumulator register A, can be broken down as follows: [note: (x) will have already been assigned a machine code value corresponding to a location in memory.]

FETCH data at (x):

1. READ instruction into IR and decode.
2. READ address word into MAR (the nature of the instruction word tells the logic control that the next word is an address, not an instruction).
3. READ data from address into SDR.

ADD data to ACC:

4. LOAD data in SDR, via bus, into ALU.
5. ADD data in ALU to data in ACC, putting result in ACC.

A typical high-level language instruction which would include this assembly level instruction is

$$z = x + y$$

Two other assembly instructions would be required, in addition to the one above in order to implement this. First it would be necessary to FETCH the data stored in address (y) and put it into the accumulator register in the processor:

FETCH A (y)

FETCH data at (y) into accumulator:

1. READ instruction into IR and decode.
2. READ address word into MAR (the nature of the instruction word tells the logic control that the next word is an address, not an instruction).
3. READ data from address into ACC.

The second assembly instruction would be that described above.

Last it would be necessary to WRITE the resulting data from the accumulator register to an address (z):

STORE A (z)

STORE data in accumulator at (z):

1. READ instruction into IR and decode.
2. READ address word into MAR (the nature of the instruction word tells the logic control that the next word is an address, not an instruction).
3. WRITE data from ACC to address.

The sequence of these three assembly level instructions constitutes an assembly level program which will achieve the intention of one high level instruction.


## 13.4 THE PROCESSOR INSTRUCTION SET

The processor control unit contains the logic to decode and carry out any one of the assembly level instructions in the 'instruction set'. These instructions can be classified into two groups, and the logic recognises which type of instruction it is being asked to decode.

1. *Data transfer instructions:*
   These relate to 'reading' and 'writing' data from/to memory locations outside the processor. Before the processor can carry out such an instruction, however, an address must be provided. For a 'direct address' instruction the instruction word is followed by a second word which gives the location in memory of the second piece of data: the address word. The word which follows an 'indirect address' instruction represents the address at which the address of the data can be found. Finally the 'relative address' instruction is not followed by an address at all; the processor takes the data from the next address in an 'instruction register'.
2. *Arithmetic and logical instructions:*
   These are concerned with operations on data some of which may already be in the processor. These instructions therefore need not always be followed by an address word.

Having decoded the instruction, the processor 'knows' whether the next word is an address or the next instruction and acts accordingly. The instruction set will also include other instructions which may be considered as aids to the programmer working at 'machine' level. These include 'increment', 'decrement' or 'skip on zero', instructions which are used in counting 'loop iterations' when running certain types of software, 'stack' instructions and 'jump' instructions which allow the processor to leave a sequence of instructions (relative addressing) and literally jump backwards or forwards in the program.

## 13.5 HIGH-LEVEL LANGUAGES

It is important to understand that 'high-level' languages consist of sets of instructions, each of which does much more than a single instruction from the processor instruction set. In fact each instruction in a high-level language does something which is much more easily understood by the ordinary person. Each instruction invokes a program of assembly level instructions, so it can be seen that a high-level language actually consists of a set of such programs. Thus a program written in a high-level language has to be 'compiled'; a process which strings together all the individual programs of assembly level instructions, each representing a single 'high-level' instruction, to produce a program comprising a list of instructions which the processor is able to execute. The process of compilation is carried out by the processor using a special program of assembly level instructions, called a compiler. Execution of the compiler program, which treats the high-level instructions as 'data', translates 'source code' into assembly instructions — 'object code'. The 'object' code is in mnemonic form and must be translated into machine code before it can be executed. This translation process, like compilation, is carried out by another special machine level program known as an assembler.

Both compilers and assemblers are special programs which must produce assembly code (compiler) and then machine code (assembler) which can be executed by a particular processor. They are therefore peculiar to one processor. Normally compilation and assembly are carried out on the same processor (or an identical one) as that on which the machine-level object code produced is to be run. However it is sometimes necessary to compile and/or assemble on a different processor: this can be done using a 'cross compiler' or 'cross assembler' program, if one is available.

## 13.6 INTERACTIVE LANGUAGES

A program of instructions, written in the first place in a high level language, is compiled and then assembled into a corresponding program of machine level instructions before being executed by the processor. The data used in execution to provide specific values for variables used in the program, is entered separately from the program itself, and can be changed between subsequent executions. Thus, for instance, a program written to execute a control algorithm would be executed over and over again at fixed intervals, each time operating on a new set of data — measured values.

In certain situations it is necessary that each high-level instruction is executed as it is translated. These situations occur when a human operator has to make inputs to the process being carried out by the processor: for instance programs written for an operator interface which allow the process operator to change set points or control actions to other values. In such cases the program may be compiled and assembled 'line by line' from the high-level instructions, allowing the program to 'write' questions to the operator and wait for him to type replies on a keyboard. Obviously such 'interactive' operation is much slower than execution of a fully compiled and assembled program which runs without the constraint of human interaction; in fact there is no comparison between the speed of operation of these two types of program.

## 13.7 'SOFTWARE ENVIRONMENTS'

In order to make it easy for a human operator to get the processor to carry out specific tasks, a special sort of high-level language is often produced which comprises a set of 'instructions' which each comprise a small program even at the normal high level. This set is automatically 'loaded' into memory and each instruction can be invoked by an operator typing the appropriate mnemonic code together with any necessary data. A good example of this is the MOS or DOS 'environment' supplied with 'IBM-compatible personal computers'. Another example is the specialised 'control language' produced by makers of distributed instrumentation systems. For a certain specialised area of operation the specialised but powerful instructions thus made available to the user make it much easier for him to configure his own system than would be the case if he had to write instructions in a generalised high-level language. The provision of such a set of special commands is referred to as providing a 'software environment' for the particular specialised use that is envisaged for the equipment. Programs written in these very specialised forms are extremely slow to execute compared to corresponding programs written in a general high-level language. Thus, the configuration of a control loop and the setting of control action values, which are functions which the user of equipment will want to change from time to time, will be written in such a form. On the other hand the control algorithms themselves, which the user will not normally wish to change, and which would be very slow to execute if written in such a form, will be written by the equipment vendor in a generalised high-level language (such as 'C' or Pascal) or even in assembler language.

# CHAPTER 14

# *Components of a Microprocessor System*

## 14.1 THE BUSES

The bus is not really a component of the system, but a 'highway' between the component 'chips'. It is actually three buses, each either 8 or 16 bits 'wide', that is to say that each bus consists of either 8 or 16 conductors so that a byte or a word can be transferred simultaneously (in parallel). In this way a separate highway is established for data, address and control. A separate highway is not required for instructions as these are *never* transferred into or out of the processor at the same time as data. A common data/instruction bus extends beyond the processor as it has to be possible to read and write to/from outside locations in memory as well as within the processor itself. Similarly both the control and address buses are extended outside the processor itself; however the processor only 'writes' addresses and so the address bus is unidirectional, which simplifies the buffers. For reasons which will be discussed later the control bus is not bidirectional either.

The address bus must provide the capability for the processor to address an adequate number of 'locations' in memory. If it is only 8 bits wide, only 2 to the power 8 = 256 locations can be addressed in a simple transfer (there are ways around this problem but they result in the processing being slowed down). On the other hand a 16-bit bus allows 2 to the power 16 = 65 536 unique addresses to be accessed by a single transfer.

One of the developments which made the microprocessor possible is the 'tri-state gate', which is an electronic logic circuit which enables data to pass onto a bus from a connected chip ('write') or from the bus to the chip ('read') or for the chip to be isolated from the bus altogether (three states). Each connection between chip and bus is 'buffered' by 8 or 16 such tri-state gates, or, in the case of a unidirectional bus, by much simpler two-state gates. Buffers have a second function: each connection makes a small power demand in order to drive the circuitry of the connected device (e.g. memory locations). The total power drain is beyond the capability of the processor to supply; the much smaller power drain to switch tri-state gates is sustainable by the processor. Power to drive the circuitry of the connected devices, provided by a separate power supply unit, is switched on/off by the tri-state gates which therefore overcome the problem which used to be posed by the 'fan out' of the processor connections to outside devices, particularly memory units.

## 14.2 THE PROCESSOR CHIP

The processor chip itself (Fig. 14.1) must connect to all three buses — data, address and control. The data bus must be bidirectional and may

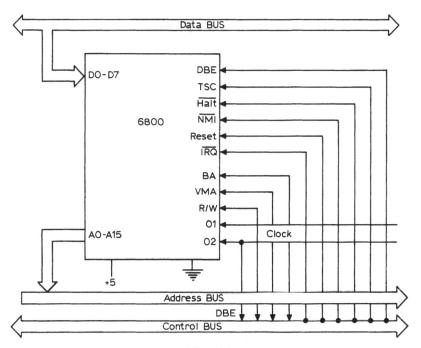

**Fig. 14.1.**

be 8 or 16 bits 'wide': the address bus, which need only be unidirectional, will probably have 16 lines to enable an adequate amount of memory to be 'directly addressed'. The control bus will comprise some lines which carry signals from the processor to other component chips, and also some lines to carry signals from other components to the processor: tri-state gate buffers are not therefore required in the control bus. The control bus will not necessarily be 8 or 16 bits wide, but will consist of as many lines as are needed to meet the design specification of the processor, and constrained by the maximum number of connecting pins it is possible to incorporate on any chip.

Certain control lines are essential to any processor:

1. 'Wait'/'halt'/'hold': a line which brings a signal to the processor from outside the sub-system, which makes it possible for a slower device or sub-system to synchronise data transfers with the processor.
2. 'IRQ' or interrupt request: a line bringing a signal into the processor from outside the sub-system which 'requests' the processor to suspend its current operation and attend to the priorities of the outside device (i.e. to read or write data from/to it).
3. R/W or read/write: a line taking a signal from the processor to set the direction of a data transfer (i.e. read or write).
4. Reset: as its name implies, a line bringing a signal into the processor to initialise its operation.

Other control lines are invariably provided, but they differ from manufacturer to manufacturer even though the purpose may be the same. For instance one may provide a signal to acknowledge 'hold' whilst the same purpose is served in another manufacturer's product by two signals: VMA or 'valid memory address' and BA or 'bus available'.


## 14.3 MEMORY CHIPS

Both ROM and RAM memory is packaged in chips. The chips vary considerably in size to suit the different requirements of individual system designers. Assuming that each 'byte' of 8 bits is allocated a unique address, 1 kbyte (1000 bytes) will require at least 1000 unique codes; i.e. 2 to the power $n$ must be greater than 1000, where $n$ is the number of bits in each address. Thus $n$ must be at least 10 (note that this already precludes the use of an 8-bit bus for direct addressing). The logic to decode the address is normally included in the chip circuitry along with the memory itself. In the case of a 16-bit address bus the remaining 6 bits can be used to select between 6 such 1 kbyte memory chips, giving a total of 6 kbytes of directly addressable memory. Alternatively, using a separate 'decoding' chip containing logic circuitry to decode a 16-bit address, 2 to the power 16 or 65 536 unique memory locations can be directly addressed. The same result can be achieved by two successive 8-bit words using a decoding chip. The first word will select one of up to 256 memory chips, whilst the second word will be decoded by this chip to select one of its 256 memory locations. Operation is considerably slower than with single (16-bit) word addressing. This is the main reason why 16-bit processors are replacing the older 8-bit processors as technological developments make them relatively cheap. An enormous quantity of memory can be directly addressed by a 32-bit word!

RAM memory is available in various sizes, starting with 256 byte chips (probably a hangover from 8-bit microprocessor days), but ROM memory, which has not been available for so long is only available in multiples of 1 kbyte.


## 14.4 THE I/O PORT

The type of interface with other systems, which any sub-system will have, depends upon the system functional specification, the nature of the communication links between sub-systems (which in turn will depend upon distance apart, security considerations and environmental factors) and the nature and quantity of information which will be transferred. The requirements will be very different for, say, an emergency shut-down system or a pipeline despatching system.

Data communication over a few meters distance is implemented in parallel mode (8 or 16 bits at a time) and is, as a result, fast. The simplest interface is the parallel I/O port which, as its name implies, acts as a 'gateway' for data to leave and enter the sub-system a byte or word at a time.

Conventional computers have different formats for addressing memory within the computer system itself and external 'devices' of any sort. This distinction has been abandoned in the majority of micro-processors and the I/O port is treated like any other address. This has the disadvantage that I/O addressing is slower, because a full address word must be generated to address the port (which is not necessary in the case of conventional computers). On the other hand it means that operations can be carried out using data from the port directly, whereas such data has to be transferred into the processor 'registers' first in a conventional computer.

Data is 'read' or 'written' by the processor in parallel, i.e. all 8 or 16 bits are 'clocked' onto or off the bus simultaneously. A port must therefore have a 'buffer' 8- or 16-bit register to store the data in as it comes in or before it goes out. In addition there must be, within the 'port' chip, facilities to 'latch' the data until transferred, means to tell the processor when data is transferred and an address decoder. Figure 14.2 illustrates these features.

If an external device wishes to write data to the processor, it will signify this requirement through an 'interrupt request' line (we will deal with interrupt requests later). At the same time it will present the first word or byte of the data to be transmitted, to the INPUT LATCH via the input bus. At an appropriate point in the cycle of operations, the processor's control unit will initiate the data transfer operation by generating the port's address followed by an INPUT STROBE pulse which 'enables' the latch, thus transferring the first word of data into the buffer. This strobe also resets the INPUT STATUS FLIP-FLOP to tell the external device that the input buffer is now full. Next the processor generates a READ signal, which, in conjunction with the ADDRESS DECODER OUTPUT, 'enables' the buffer, causing data to be read into

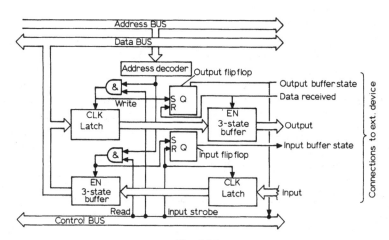

**Fig. 14.2.**

the processor, and leaving the buffer empty once again. The same signal that enables the buffer, also resets the STATUS FLIP-FLOP to indicate 'buffer empty' to the external device. Between the time that the external device receives the 'buffer full' and 'buffer empty' signals it must present the next byte or word of data to the latch on the input bus.

Similarly, for the processor to write to an external device, it must alert that device and present the first word or byte, via the data bus, to the OUTPUT LATCH. Next the processor addresses the output port and then generates a WRITE SIGNAL which, in conjunction with the ADDRESS DECODER OUTPUT, causes the data to be transferred to the buffer, at the same time setting the output status flop-flop to 'buffer full'. Thus the external device knows that data is available. However, on this occasion, it is the external device which generates the signal to initiate data transfer. This 'data received' signal also sets the buffer status flip-flop to the 'buffer empty' state. The processor must present the next word/byte to the OUTPUT LATCH between the time that 'buffer full' and 'buffer empty' states are signalled.

These operations are known as 'hand-shaking' as control of the operation (of data transfer) is split between the two sub-systems involved.


## 14.5 A BASIC MICROPROCESSOR SYSTEM

It is now possible to see how a rudimentary microprocessor system is constructed from the elements discussed so far (see Fig. 14.3). The system comprises a 16-bit processor chip with up to 4 ROM chips, up to 8 RAM chips and a 2-port chip. The six highest-order address lines A10–A15 are connected to 'chip select' (CS) pins on the ROM, RAM or port chips. Two of these CS lines, A13 & A14, are connected to ROM chips, enabling the processor to select from a maximum of 4 (2 to the power 2) ROM chips. Similarly the three CS lines A10–A12 are used to select one of the 8 (2 to the power 3) RAM chips. Having selected the chip to be addressed, the lines A0–A9 are decoded by the chip selected as the address of the data which is to be 'read' or 'written'. Lines A0–A9 are connected to all the RAM & ROM chips, so that each can have a maximum capacity of 2 to the power 10 = 1 kbyte. The remaining line, A15, is used to select one or other of the two ports. Thus this system does not need to use an address decoding chip which would be required if any more ROM or RAM memory, or more ports were included.

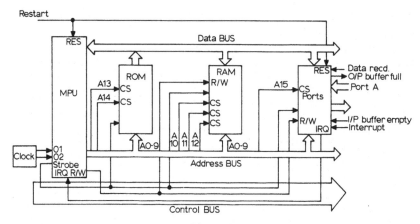

**Fig. 14.3.**

## 14.6 SERIAL I/O INTERFACING

Data transfer over distances much greater than a few metres is likely to suffer degradation of and interference to the signals. This subject is dealt with in sections 16.3 and 16.4 but it is a fact that serial (bit-by-bit) transmission is essential beyond a few metres. Some form of interface is necessary therefore to enable the system to convert parallel data to serial form. Any such interface will have to incorporate 'buffer' storage registers to accommodate the parallel words or bytes of data, as well as the means to control the communication process. Figure 14.4 shows the pin connections to a typical communication interface adapter or CIA. The left-hand side is similar to the port chip. The right-hand side however has a single transmit (TXD) line and a single receive (RXD) line in place of the parallel bus connections to the port chip. It also has two 'clock' connections TXC and RXC which allow the data transmission speed to be controlled by either the system processor or the external device (which may be another processor in a different sub-system). Thus the slower device can be allowed to control speed of transfer. Typically the interface chip will also incorporate several connections to a 'modem'. This is a device which is used to control data transfer over long-distance communication links; its functions include error detection. The CIA is therefore a device which enables the processor to interface with other processors via simple communication links, such as a twisted pair of conductors, or via a complex telephone or radio communication link.

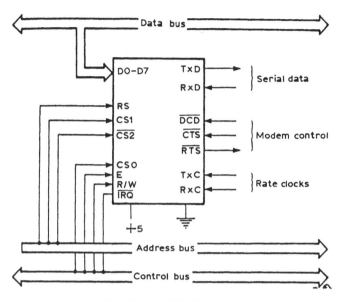

**Fig. 14.4.** ACIA functions.

## 14.7 INPUT/OUTPUT CONTROL STRATEGIES

The structure of a microprocessor system is determined by the way that each processor communicates with others. In the simplest systems an 'embedded' processor, controlling say a washing machine, and operating from a fixed program held in ROM, will 'pole' a number of switches in strict rotation on a continuous basis. 'Polling' consists of determining the state of the switch, ('open' or 'closed' — '1' or '0'). The system will only require the processor, a clock, a single port, some ROM and some simple 'driver' circuitry to interface with the switches. Such switches will be attached to level or temperature sensing detectors and will tell the processor when water level or temperature reach preset limiting values. The processor will also 'write' or output '0' or '1' data to devices such as contacters for pumps, solenoid valves or heaters to control the sequence and timing of events during the machine's operating cycle. At the other end of the spectrum a large 'emergency shut-down system' on an oil rig, for instance, can be constructed using the same microprocessors but will comprise many, much more complex sub-systems which will have to communicate with each other over some distance in many cases. The processing of data in such a system is distributed according to functional and geographical constraints. Polling is still the most common method used to obtain information from sensors, but a system of priorities will be needed to ensure that urgent data is transmitted in preference to less urgent. Priority decoding chips are employed for this purpose: each element in the system can 'raise a flag' (set a special '1'/'0' bit) to indicate that it is ready to write data to a processor and the decoder chip will initiate the processor interrupt on a priority basis. A status decoder chip may be used to speed up this operation by providing the processor with the starting address in memory appropriate to servicing the device which is causing the interrupt: the processor would thus be saved many operations involved in searching its own memory to find this address.

Interrupt facilities, priority decoding and status decoding chips can overcome the problem of fast access by one processor in a system to another, but there remains the problem of fast 'servicing' of the interrupt. Whenever an interrupt occurs in a processor's operation, the processor must carry out a 'housekeeping' routine to store the data, addresses and instructions it is currently working on so that it can return to them when the interrupt has been serviced. Most interrupts occur in order that an external sub-system can transfer data to the processor's memory at a particular time (for instance, an external system may have the function of 'polling' a number of measuring instruments and then transferring the measured-value data to another sub-system at regular intervals of time). In such cases the answer is once again to add to the system a specialised (function) chip, in this case a 'direct memory access' or DMA chip.

## 14.8 DIRECT MEMORY ACCESS

The direct memory access chip is typical of the most complex of the VLSI chips being manufactured as 'hardware building blocks', from which the system builder can design and construct very powerful computing systems (as opposed to computers). The DMA chip in its most complex form can be regarded as a specialised processor in its own right.

The simpler DMA chips operate through the interrupt line to suspend the operation of the processor they are associated with in order to 'steal' processing time for the direct transfer of data from another processor into memory. They speed up operation for two reasons:

1. The processor does not have to carry out any 'housekeeping' routine, but only stop its operations while the transfer takes place.
2. The transfer of data is controlled by a program held in the 'firmware' of the DMA chip rather than software which is part of the operating system of the processor. Software is always slower in operation than 'firmware'.

Nevertheless data transfers obviously still slow down the processing operations carried out by the processor because these have to be stopped during the transfer operations. If large amounts of data must be frequently transferred, such delays may be intolerable. The operation of this type of DMA chip is illustrated in Fig. 14.5.

More sophisticated DMA chips make use of the fact that the processor does not need to use the address and data buses when carrying out some of the instructions in the processor instruction set. Thus the DMA chip can utilise these buses in intervals when the processor is carrying out a 'processing' rather than a 'read or write' instruction to transfer at least part of the data. Such chips are very expensive, however, because they only sell into a limited market. In order that a DMA chip can be incorporated into a system, the processor must provide several additional control lines, and tri-state buffers must be incorporated into the address and data buses. These additional control signals include some form of acknowledge signal to the DMA to

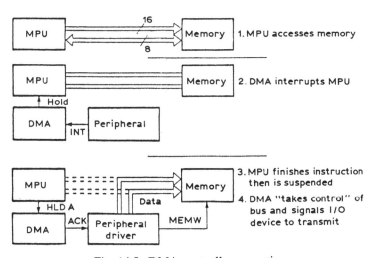

**Fig. 14.5.** DMA controller operation.

confirm that the processor has received and acted upon a HOLD or WAIT signal from the DMA chip. This tells it that the data and address buses are available for its use. Another additional control signal, generated by the DMA chip, tells the processor to set the tri-state gates so that the DMA rather than the processor can use the buses.

# CHAPTER 15

# *Data Acquisition Sub-Systems*

## 15.1 DATA ACQUISITION

One of the most important functions of an instrumentation system, and one which is not required of other types of system, is the gathering of measurement data, at regular time intervals, from, often, large numbers of individual measurement transducers. This requires 'real-time' techniques which are not found in other systems and therefore chips which are unique to such systems. Measurement data is generated by individual transducers continuously; however if many such measurements are to be 'read' by the computing system, each must be sampled at regular intervals in turn and the data reconstructed in the system from these samples. The frequency of the sampling must reflect the way in which the measured values themselves change with time. (A rapidly changing measured value will have to be sampled much more frequently than a slowly changing value if it is to be possible to reconstruct its true nature from the samples (Fig. 15.1).)

There are several possible '*modi operandi*' for this data sampling and transfer process:

1. Groups of transducers may be 'scanned' in strict sequence by a sub-system, the data then being organised into a 'data base' from which 'packages' of data are constructed and transferred to a central system to be incorporated into its data base. The frequency at which such 'packages' would have to be transferred would depend upon the nature of the most rapidly changing measured value in the group. The more frequently such 'packages' must be transferred within the system, the fewer

275

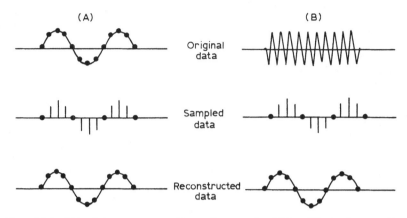

**Fig. 15.1.** (A) Adequate sampling (8 samples/Hz); (B) bad sampling
(> 2 samples/Hz).

measurements could be included in the sub-system, depending
on the time it takes to 'package' and transfer a set of values.
2. Any one of a group of transducers may be 'addressed' by the
processor and a sample of its measured value transferred to one of
the processor's internal registers for immediate processing.
3. Some combination of methods 1 and 2 above. Some of a group of
transducers can be 'polled' on a regular basis, whilst the
remainder can be addressed by the processor when required.
Alternatively some of the group may be 'polled' only at every
second or fifth interval, say, whilst the remainder are 'polled' at
every interval.

Which of these options is adopted in the design of a 'data acquisition'
sub-system will depend upon the design philosophy of the instrumentation system. Often more than one such option is incorporated into the
design of the sub-system in order that the design of the instrumentation
system can be 'tailored' to suit a user's requirements.

Regularly 'polled' measured values are 'packaged' at the end of the
polling cycle and the resulting data package is then transferred to the
data base in another sub-systems memory, often by direct memory
transfer. The advantages in terms of handling data are obvious; the
processor in the receiving sub-system need not address the data until it
is ready to. The penalty is that the 'age' of the data is not accurately
known because the operations of gathering, packaging and transferring
the data are independent of its processing. Provided that the frequency

of polling is high enough, this uncertainty of 'age' may not be significant: the frequency of polling, in turn, determines the number of transducers which can be included in any one group. Method three above allows some flexibility to include both fast-changing and slower-changing measurements in a group and at the same time maximise the number of transducers. In a very large instrumentation system, however, it may be better to have two separate data acquisition sub-systems (or more) rather than mix 'fast' and 'slow' measurements.

Method 2 above, in which the gathering and transfer of measured values is not independent of processing, both being controlled by the processor of the receiving sub-system, overcomes the 'age' problem. The penalty is in overall speed of operation, and in large systems this penalty is usually too severe to be acceptable.

A very common system structure utilises several 'free-running' (method 1) sub-systems to scan a limited number of measurement transducers and to regularly transfer packages of data to a central sub-system for processing. This structure allows the numbers and types of transducer (providing analog, discrete (1 bit — on/off) or digitally coded signals) to be distributed between the several data-acquisition sub-systems: one might scan a large number of slowly changing signals, another a large number of discrete signals and yet another a small number of fast-changing or alarm-critical signals. In addition there may or may not be included a sub-system which allows a limited number of signals (of all types) to be addressed directly by the processor of the central sub-system as and when it required such data.

Each data acquisition unit will comprise, in addition to the basic components of any sub-system (as described in Section 14.5), a number of 'specialised' chips which will provide the following functions:

signal multiplexing
'sample and hold'
analog/digital signal conversion
signal conditioning

## 15.2 THE SIGNAL MULTIPLEXER

A multiplexer is a device which enables a number of signals to share a single communication channel. It achieves this by switching each signal in turn to the communication channel for a fixed interval of time.

Implicit in this *modus operandi* is the need to 'sample' the signal at a specific instant and to 'hold' the sampled value constant while the data transfer is carried out. When a multiplexer is incorporated into a data-acquisition sub-system the data transfer may be postponed until a package of data has been assembled, but this does not remove the need to sample and hold the individual signals. If the signals are in analog form then the hold operation must be extended to the time required to convert the sampled analog value into digitally coded form. The multiplexer (MUX), the 'sample and hold' (S/H) unit and analog-to-digital signal converter (A/D) may all be implemented as separate chips, or may, in small systems, be incorporated into one single chip.

The multiplexer must incorporate an address decoder and control logic, so that it can either respond to a direct instruction from a processor to sample a particular channel or can 'scan' each signal in turn continuously, or on demand. The control logic must provide that the sample and hold functions synchronise with the A/D function. The switching unit itself must be designed to meet the particular needs of the instrument system: for instance if thermocouple or RTB signals are to be switched, then switches will have to be specially designed for low contact resistance and double-pole switching may well be required. More often however analog signals will be generated by instrument transducers in the form of 4–20 mA current loops. Discrete signals will probably be in voltage form, and provide at least $\pm 5$ V. For such signals, switching resistance is not usually of critical importance.

## 15.3 ANALOG-TO-DIGITAL CONVERSION

Conversion of analog measurement signals into discrete digitally coded values is fundamental to the application of digital computer techniques to industrial instrumentation systems. Because digital values are discrete and not, as analog values, continuous, conversion requires that a search is made for the closest possible correspondence between the analog value and its digital equivalent. A voltage proportional to the discrete value of each bit is generated and added together to represent the analog equivalent of any given binary number. Thus for an 8-bit A/D converter:

For the least significant bit (LSB) voltage = (2 to the power 0) $\times V = V$

For next most significant bit voltage = (2 to the power 1) × $V = 2V$

$$\vdots \qquad \vdots \qquad \vdots \qquad \vdots \qquad \vdots$$

For the most significant bit voltage = (2 to the power 7) × $V = 128V$

The scaling factor $V$ can be evaluated by setting the maximum value of the analog signal equal to the maximum of the digital representation. For instance, if the analog range is 0–10 volts, then

10 volts = the sum of $(128V + 64V + 32V + 16V + 8V + 4V + 2V + V) = 255V$

Hence

$$V = 10/255 = 0{\cdot}0392 \text{ volts}$$

Thus the least significant bit represents $0{\cdot}0392$ volts whilst the most significant bit represents $5{\cdot}0200$ volts.

By systematic trial and error the voltage equivalent of digital values are compared sequentially with the actual measured value until the error between them is less than the voltage value of the least significant bit (see Fig. 15.2).

This provides the closest correspondence of the digital-to-analog measured value which is possible using 8 bits, and is taken as the digital equivalent. A maximum error of half the value of the least significant

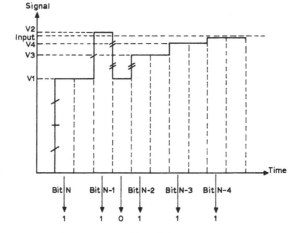

**Fig. 15.2.**

bit, taken as a fraction of the maximum of the measurement range, is implicit in this conversion. If, therefore, a resolution of better than 1 in 1000 is required, a minimum of 10 bits must be used. The method is illustrated by the following table:

| Analog value | Binary equivalent | 'Bit' voltages |
|---|---|---|
| | 0 (MSB) | 0 |
| | 1 | 2·509 8 |
| | 0 | 0 |
| 3·6 volts | 1 | 0·627 4 |
| | 1 | 0·313 7 |
| | 1 | 0·156 9 |
| | 0 | 0 |
| | 1 (LSB) | 0·392 2 |
| | Total = 4·000 0 volts | |

The error band = LSB = 0·3922, so that the greatest error possible due to conversion is ±0·1961 volts; in this case however the error is only 0·0078 volts.

The nearest equivalent digitally coded binary value is therefore 01011101, which can be transmitted as a series of '0' or '1' pulses.

The error band, and therefore the precision of the conversion depends upon the number of bits used; the greater the number of bits, the greater the precision. However, the greater the number of bits, the longer the process of trial and error takes to arrive at the digital equivalent value. Selection of the type of A/D converter therefore depends upon the relative importance of precision and speed for the particular sub-system concerned. The time taken for conversion is minimised by a systematic search technique: the first 'guess' is taken to be mid-range, then depending on whether the comparison shows that the measured value is greater or less than this, the next guess is taken as the mid-value in either the upper or lower half of the range. Subsequent guesses similarly halve the remaining range of uncertainty until the closest approximation is isolated. Using this technique the number of attempts required to find the digital equivalent are log to the base 2 of 2 to the $(n - 1)$, where $n$ is the number of bits used in the conversion. For a 12-bit converter, therefore, this works out to 11 attempts which is not unreasonable. Thus precision and speed are not necessarily incompatible. If very high speed is required, however, parallel conversion can be used. With this method the voltage equivalents of all possible digital values are made available simultaneously and the actual value compared simultaneously with each. Parallel logic detects at once which comparisons give a positive error and which a negative error, and thus immediately isolates the best digital approximation. The circuitry required to implement this method is very considerable, and it is rarely used at present.

## 15.4 SAMPLE AND HOLD CIRCUITRY

However quickly the analog-to-digital conversion is accomplished, it may take a significant time in respect of changing measured values. The measured value must therefore be sampled and the sample value frozen for sufficient time to allow the conversion to be carried out. This is accomplished by allowing an electrical capacitor to charge up to the analog signal voltage: however this too takes a significant time. S/H circuits can be placed before or after the multiplexer, but if before, they must be provided separately for each measurement signal (i.e. there will be as many sample and hold circuits as there are measured value signals). If sampling is carried out after multiplexing, however, only one S/H circuit is required. However, in this case the sampling delay must be added to the conversion delay because the conversion process must wait until a sample has been established before it can proceed, whereas if multiple samplers are provided in front of the multiplexer, sampling can be timed to start before the multiplexer scan has reached the relevant measured value. Thus for fast-changing measurements the design of the data acquisition sub-system may incorporate a sample-and-hold circuit for each measured value 'input'.

## 15.5 SAMPLING

The purpose of industrial measuring systems is almost always to show the way a measured variable varies in time. This is why, until the appearance of digital computation in instrumentation systems, measured values were transmitted as analog signals in mechanical (linkages), pneumatic, electrical or hydraulic form. If these *continuous* analog signals are to be sampled and subsequently represented by *discrete* values of this sample, then it must be possible to reconstruct the original continuous signal for display or other purposes from these discrete sampled data. The question which inevitably arises, therefore, is 'how frequently must samples of any particular measured value be taken in order that its true behaviour can be reconstructed from the samples?' The well-known 'sampling theorem' states that the frequency of sampling must be at least twice that of the highest frequency component of the measured-value time function. This is a deceptively simple answer which depends upon the principles of Fourier analysis, that is, on the fact that *all* time-variant functions can be 'decomposed' into a number of sinusoidal functions of varying frequency and amplitude. This is not difficult to understand; it is well known that any musical cord can be 'decomposed' into its constituent 'pure notes' which are in fact sinusoidal functions of sound pressure as sensed by our ears. It can readily be appreciated from the example illustrated in Fig. 15.1 that, whilst the first function can be faithfully reconstructed from the samples taken, the second cannot. Both signal functions are sinusoids, but whereas in the first case (A) the frequency of sampling is many times the frequency of the signal, in the second case (B) the frequency of sampling is just slightly less than that of the signal, giving rise to a reconstructed 'signal' which is actually the 'alias' of the real signal. The alias in this case, where the real signal is a sinusoid itself for simplicity, is the sinusoid of frequency, equal to the difference between the real signal and the sampling frequency. The difficult part is to find out what the highest-frequency sinusoidal component of a real signal is! In practice, however, the sampling rate is usually fixed at many times this value to avoid any problem.

## 15.6 DATA 'PACKAGING'

In addition to the basic functions outlined earlier, and the signal processing, multiplexing, sampling and A/D conversion functions a data acquisition sub-system will require the means to package sets of measured data for transmission to a central processing sub-system once every multiplexer scan. A processor will probably be used to carry out this function completely separately from the other functions, the data being stored in buffer memory as produced. The functions involved in data transfer over different types of communication channels is the subject matter of the next chapter.

## 15.7 ARCHITECTURE OF A DATA-ACQUISITION SUB-SYSTEM

The basic microprocessor system presented in Section 14.5 can now be developed into a typical data-acquisition sub-system, as illustrated in Fig. 15.3. Whatever the function of the sub-system may be, data transfer to other or a central unit will almost invariably have to be in serial form over a proper communication channel. This is the function of the CIC or ACIC (Asynchronous Communication Interface Chip), which is shown in Fig. 15.3 communicating with other sub-systems via a single pair of conductors. Inside the sub-system all component chips are connected by 8- or 16-bit buses. Often all the chips will be mounted on a single 'motherboard' with the buses laid out at the manufacturing stage. Sometimes, however, some components will be mounted on a second (or third or more) circuit board and the circuit boards interconnected by 'ribbon' 8- or 16-wire cables. Connections with other sub-systems, which may not necessarily have been made by the same manufacturer, will necessitate following standard practices with regard to connectors, physical factors such as voltage levels of signals and the protocol of data transfers.

The component parts of a typical data-acquisition sub-system are shown in block-diagram form in Fig. 15.3. An address decoding chip

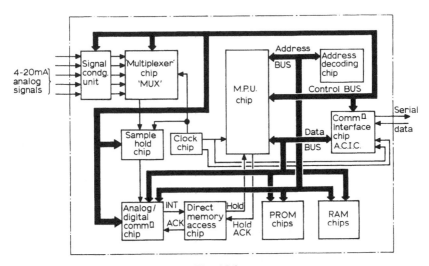

**Fig. 15.3.**

will probably be included in order to speed up access by the processor chip to fairly large amounts of memory. Almost certainly a direct memory access chip will be used because of the large data-transfer flow rate implicit in measured-value sampling constraints. In addition there will be signal conditioning, multiplexing, S/H and A/D chips. In a large system a second processor might be used for data packaging and transmission control, leaving the first processor to control the scanning or addressing of input data asynchronously. This means that the operations of gathering the data form the measurement transducers, and the operations of packaging and transmitting the data run independently in their own time; there is no need for them to be synchronised. However there is as always a penalty to be paid for this independence in as much as the data taken from memory for packaging is then of uncertain age. Operation of a system having only one processor must be synchronised to the processor's clock, thus avoiding the data age problem.

# CHAPTER 16

# *Data Transfer Between Sub-Systems*

## 16.1 INTRODUCTION

Invariably a sub-system will be required to send data to and receive data from at least one and often many other sub-systems within a complete system. Whilst the communication links between elements within a sub-system (the bus) are in a protected environment, the communication links between sub-systems will have to fulfil their role within an industrial environment which is often extremely 'hostile'. This 'hostility' derives from the electrostatic and electromagnetic fields set up by large electrical rotating machinery and switchgear, as well as 'stray' electrical leakage currents (often many times greater than the signal currents) and, even more recently, high-powered radio or radar signals. Long lengths of metal conductors (signal cables) present considerable distributed resistance, capacitance and inductance in the path of such 'hostile' electrical energy so that there is every likelihood that some of it will be transferred into the communication channel leading to malfunctioning, or more often error data being included in data packages. The defences against these problems include techniques for preventing such energy transfer, together with techniques for detecting errors that do occur and for eliminating these.

There are however other considerable advantages which stem directly from the adoption of discrete sampled data representation of measured-value data. These include the display flexibility and cost savings attendant on the substitution of the now familiar visual display units (VDU), the facility for 'historical data' recording and play-back and data security. The last advantage stems from the opportunity to provide back-up communication channels for *all* communications, often following a different geographical route, which is clearly not on

285

with the old 'dedicated pair' links. Other very considerable advantages stem from the adoption of digital computational techniques for control mechanisms in place of the previous analog techniques. These advantages are at least as important as the communication advantages and will be dealt with later, in the proper context.

## 16.2  DIGITAL SIGNALLING

The transmission of analog measurement signals over electrical or pneumatic channels is subject to delay due to the capacitance and resistive characteristics of the communication link itself. There is always the possibility that, in addition, electrical current (or air) leaks or unwanted 'pick up' will degrade the signal further. The advantages of digitally coded values, even though these inevitably imply the use of sampling and subsequent signal reconstruction techniques, stem from the fact that a single bit of digital data is indicated by whether a '1' is or is not present at a particular time. There can be a wide tolerance on the voltage or current flow which represent '1' and, as long as presence or absence can be reliably determined within such a wide tolerance, no error at all is incurred in the data transmitted. A specific interval of time must be allocated to the generation (at the transmitting end of the communication channel) and subsequently for detection (at the receiving end). The length of this interval determines the speed at which 'bits' of data can be sent, and together with the number of 'bits' required to encode each sampled value of the measurement, the speed at which sampled measurement data can be transmitted. As with analog signals, capacitive delay caused by the transmission cable will distort the signal, in this case a 'pulse' of a magnitude dependent upon the signalling levels (voltage in most cases) and duration dependent upon the speed of transmission. The longer the cable and the greater the resistive/capacitive delay characteristics of the cable used, the greater the distortion caused to any particular pulse. An ideal pulse will consist of an instantaneous 'rise' (of voltage), an interval during which the signal (voltage) is constant and an instantaneous 'fall'. The effect of capacitive/resistive delay is to make the 'rise' and 'fall' less than instantaneous, and the extent to which this happens depends upon the length of the cable and its electrical characteristics. Beyond a certain point the duration of the pulse, which is directly proportional to the speed of transmission, becomes important also; the shorter the duration, the sooner (in terms of cable length) the magnitude of the pulse begins to diminish. These effects are illustrated in Fig. 16.1.

Data transmission over distances of 1200 m or more at transmission speeds in excess of 3000 bits/s can present difficulties even when special low-capacity cable is used. Many microprocessor systems require to work at 20 000 bits/s or 20 kbaud and so 'line drivers' are usually installed at suitable intervals in the communication cables in systems in which inter-unit distances are of this order. A line driver is a device which receives the data, detects the absence or presence of a 'pulse' within each interval, and regenerates the data with 'new' pulses so that it can survive transmission over a further distance span. Provided that line drivers are installed at intervals such that pulses can be accurately detected, there is no loss of data accuracy or integrity at all, no matter how great the transmission distance. Beyond distances of a kilometer or so the cost of line drivers becomes prohibitive and 'limited distance modems' are likely to be used (see Section 17.7).

## 16.3 SIGNAL INTERFERENCE

In industrial environments, because of large rotating electrical plant, the 'earth' or 'ground' potential (voltage) at different locations can be several volts different. This is often true even when the locations are quite close together, and the effect on communication links can be very serious indeed. The difference in potential sets up 'ground-loop' circuits as shown in Fig. 16.2 (a) in which the ground forms one conductor whilst the signal cable forms the other. The current thus induced to flow in the ground loop circuit is of an alternating nature because the 'earth leakage' currents from the rotating electrical plant are themselves of this nature. Nor is the alternating frequency simply equal to the mains generating frequency; such leakage currents comprise a wide range of frequencies, some of which may easily correspond closely to the signalling frequency or pulse repetition rate of the signal. Ground loop circuits therefore introduce the near certainty of serious interference in communication channels used in industrial environments, and must be avoided at all costs.

The problem can be solved by using separate 2-wire circuits for data transfer in each direction, as shown in Fig. 6.2(b). The equipment required to achieve this uses differential amplifiers (see Section 17.8) and is much more complex. Of more importance is the requirement for

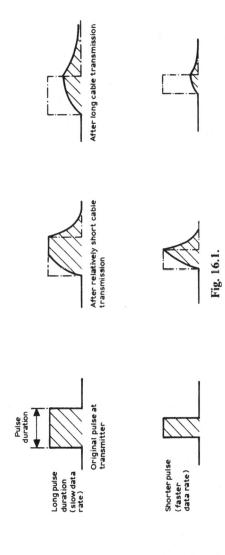

Fig. 16.1.

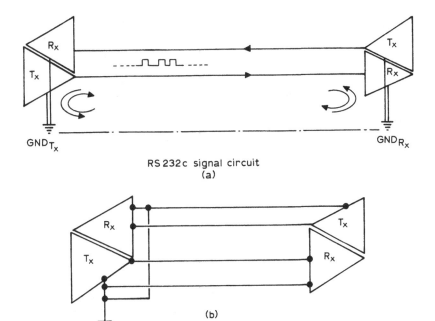

RS 232c signal circuit
(a)

(b)

**Fig. 16.2.** $R_x$ = receiver; $T_x$ = transmitter.

four rather than two conductors in the communication link cable, as this more than doubles the cost of cable.

Quite apart from the possibility of 'ground-loop' interference there is always the possibility of introducing 'common-mode' interference into the communication link by two mechanisms which are commonly encountered in industrial environments and indeed in any environment to some extent. [The term 'common mode' refers to any interference that affects both conductors in the cable equally.] The first of these mechanisms introduces energy by electromagnetic induction, at frequencies determined by the source, from power cables which run parallel and close to the communication cables in the same way that power is transferred from the primary to the secondary circuits of a transformer. The second of these mechanisms is 'electrostatic induction', in which electrical energy is transferred from a radiating source by radio transmission through the 'ether', High-frequency voltage variations are thus induced in the communication cables in the same way that radio or television signals are induced in aerials.

Protection against common-mode interference can be provided by two techniques. The first is to use 'twisted pairs' of conductors, which greatly reduces the danger of electromagnetic interference because the alternating voltages induced in successive sections formed by the twisting process will be of opposite polarity and will therefore tend to cancel each other out. Electrostatic interference can be greatly reduced by 'screening' the cables and grounding the screen so that it, rather than the signal conductors, acts as an aerial and by-passes the noise to ground.

## 16.4 FREQUENCY SHIFT KEYING

Degradation of the pulse shape by cable capacitance and resistance, up to the point when it can no longer be recognised, does not result in any loss of information: nevertheless transmission of digitally coded data comprising theoretically square voltage pulses will become increasingly error-prone as the peak voltage of the '1' state diminishes, and is reduced in extreme cases to the level of the noise or 'grass' due to interference as shown in Fig. 16.3.

Such signal degradation can be greatly reduced by a different coding technique. A '1' is represented by a burst of a sinusoidal signal of one frequency whilst a '0' is represented by a burst of another frequency as shown in Fig. 16.4. It is a property of a sinusoid that the rate of change is constant, and so the rate of change of voltage or 'slew rate' implicit in FSK signal coding is much less than the theoretically infinite rate implicit in the simpler voltage level pulse technique. The effect of the capacitive/resistive characteristics of the communication cable on an FSK signal is to cause a 'phase delay' rather than to diminish the amplitude, and so the all-important 'signal-to-noise ratio' is preserved

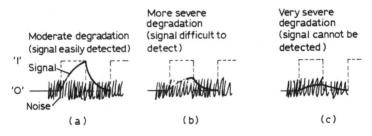

**Fig. 16.3.** Signal-to-noise ratio.

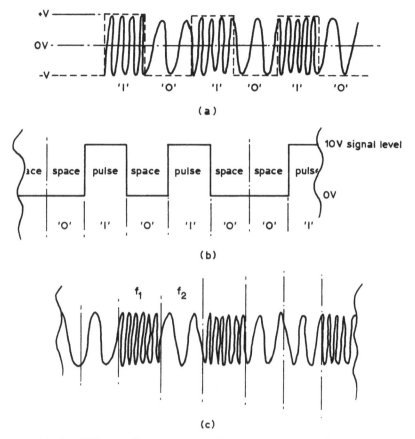

**Fig. 16.4.** (a) FSK signalling; (b) conventional signalling; (c) frequency-keyed signal.

over much longer distances. However the phase delay does have a limiting effect on the speed of data transfer in another way. The frequency of the signal representing a '1' state must be sufficiently different from that representing the '0' state that the receiving equipment can readily distinguish between them: but the phase delay suffered by each will be indeterminate and may cause timing problems at the receiver in extreme cases. This problem can be lessened by increasing the pulse duration: nevertheless this, together with the difference in frequency representing '1' and '0', will determine how short the duration of each pulse can be and will therefore limit the speed of

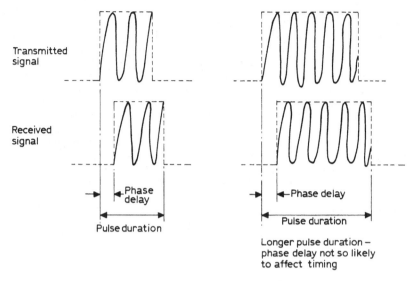

**Fig. 16.5.** Phase distortion — FSK signalling.

data transfer. The higher frequency 'keyed' will be limited by the 'band-pass' characteristics of the communication channel — the highest frequency which can be transmitted without suffering attenuation of more than 6 dB. The lower frequency will be limited by the 'bandwidth' of the channel (see Section 16.5 — FDM).

## 16.5 MULTIPLEXING OF COMMUNICATION LINKS

Unless speed of transmission is an absolute priority, long communication links are usually 'shared' between a number of transmitter/receiver pairs to reduce costs. There are basically two ways that this 'multiplexing' can be achieved, 'Time division multiplexing' (TDM), or 'Frequency division multiplexing' (FDM). The former technique depends on packaging data into 'messages' which are transmitted during a 'time slot' during the 'message repetition period'. Messages can thus only be of a finite maximum length, and the 'message repetition rate' also depends upon the number of channels utilising the communication link. FDM multiplexing depends on splitting the link into 'frequency channels' each of which is modulated separately by frequency keying to carry data continuously rather than in time slots.

The drawback with this technique is that the 'bandwidth' of each channel, and hence the maximum rate of transmission possible, is greatly reduced, not only in proportion to the number of channels but additionally because of the need to maintain adequate 'channel separation' to avoid 'cross talk'. In practice it is usually found that digital data can be transferred faster by TDM techniques than by FDM, but the message packaging techniques to achieve this are far from simple. Sampling techniques have to be used and continuous measurement signals reconstructed from these samples in the receiving equipment.

## 16.6 ERROR DETECTION

Data transferred between sub-systems of a distributed system must be protected from signal degradation as it travels along the communication link itself. For this reason data is nearly always transferred serially and in message packages: not only is every precaution taken to ensure that data is not degraded, and that error data does not enter the message, but techniques are applied in the construction of each message to detect any errors that may nevertheless occur.

The techniques of error detection and correction have been developed into a branch of mathematics which is totally beyond the scope of this book. Nevertheless an understanding of the basic principles is essential to the proper understanding of communication of data.

The fundamental principle of error detection rests on the concept of 'parity'. To any set of binary data a single pulse interval is added in which the presence or absence of a pulse ('1' or '0') ensures that the total number of '1's is either odd or even. If then, during transmission, a single bit is 'lost' or 'gained' due to any of the causes described in Section 16.3, parity will be incorrect at the receiving end and it can be deduced that an error has occurred. In many systems this is sufficient protection, and should an error be detected, the receiving sub-system 'requests' a repeat of the entire message. If, however, the environment is so hostile or unpredictably hostile that multiple bit errors are a real possibility, then simple parity checking of the entire message will not suffice since, if two or any even number of single-bit errors occur, they will not be detected.

A more complex form of parity checking, known as 'Hamming code' will serve to illustrate how multiple errors can be detected and 'single-bit' errors corrected without the need to repeat the message. For an 8-bit 'byte' of data, 4 additional bits are added for parity checking.

The 8 bits of data are grouped into threes as shown below

| | | | |
|---|---|---|---|
| bit '0' | bit '3' | bit '6' | parity bit '2' |
| bit '1' | bit '4' | bit '7' | parity bit '3' |
| bit '2' | bit '5' | | |

parity bit '0'     parity bit '1'

so that parity bit '2' checks the validity of the group of data bits '0', '3' & '6', whilst parity bit '1' does the same for the group of data bits '3', '4' & '5'. If now bit '5' were to be changed by interference (from a '1' to a '0', or from a '0' to a '1'), then the parity check of the group of data bits '3', '4' & '5' (parity bit '1') would signal an error since the total would not be even (or uneven) as expected at the receiving end. Similarly, if data bit '1' were to be corrupted, the parity check of both the groups of data bits '0', '1' & '2' (parity bit '0') and '1', '4' & '7' (parity bit '3') would signal an error and thus indicate which data bit had been corrupted. Since bits have only two possible states a corrupted bit can be corrected. Thus single-bit errors caused by occasional interference will not only be detected, but can be corrected at the receiving sub-system without the need to request a retransmission of the entire message. The penalty in terms of speed of data transfer is heavy, however; 50% increase in the number of bits, hence 50% increase in the time for transmission as against 100% approximately for retransmission. More importantly, multiple bit errors will now almost certainly be detected, though they cannot be corrected at the receiving end. The communication channel must in reality be protected adequately from all but unusual interference, since no error detection technique can correct for multiple bit errors, and repeated message repetition will bring communication more or less to a halt. In practice this is almost always a matter of the signal levels, the data transfer rate and design of the link so that interference is avoided as far as possible.

## 16.7 DATA HIGHWAYS AND COMMUNICATION NETWORKS

The simplest form of communication link is a pair of solid wire conductors dedicated to the transfer of data between two sub-systems. For systems in which there are a large number of sub-units at one or both ends of a communication link (such as data-gathering systems) multiplexing techniques can be used to avoid the cost of multiple 'dedicated' links. However for systems in which many sub-systems are distributed over a wide area, rather than concentrated at the two ends of a single communication link, and in which each must communicate with a number of others, the multiplicity of 'dedicated' pairs of conductors very soon becomes excessive, as well as being inflexible. For instance, in a system comprising just 5 sub-system units, complete flexibility (each unit being able to 'talk' to every other unit) requires a number of separate communication channels equal to factorial 5 (120), whilst for a system comprising 12 units — a relatively small system — the number is factorial 12 = 48 000 000, which is clearly not on. For this reason the advent of distributed processing initiated a revolution in the design of large data-based systems, not least amongst which is the industrial instrumentation system. For a long time the cost of multiple single pair analog communication links in large instrumentation systems for plants such as steel works, aluminium smelters, petroleum refineries, chemical plants, etc., had represented up to 20% of the total cost of such instrumentation systems, which themselves represented up to 12% of the total cost of the plant (i.e. up to 2·5% of the total cost of plant costing many millions of pounds or dollars).

The fact that the number of dedicated links in a system in which all the sub-systems 'talk' to each other becomes prohibitive when the number of sub-systems is still quite small was thus a considerable barrier to the development of interactive control systems. The development of fast digital data transfer systems has made possible the use of digitally encoded sampled data in instrumentation systems. This in turn has introduced the possibility of using data-packaging and TDM telemetry techniques in order to utilise a single common communication channel to interconnect a large number of sub-systems.

Communication channels for distributed processing systems usually take one of several common forms which are illustrated in Fig. 16.6. The 'Star' topology is adopted for systems in which a number of local or satellite sub-systems talk to and are controlled by a central sub-system, usually referred to as the 'master' unit. In such a system, data passed from one local unit to another must travel via the central unit: if there is much such data transfer required, the central unit will soon become overloaded. The degree to which data processing can be distributed in such a system is strictly limited by the lack of any direct communication between local units.

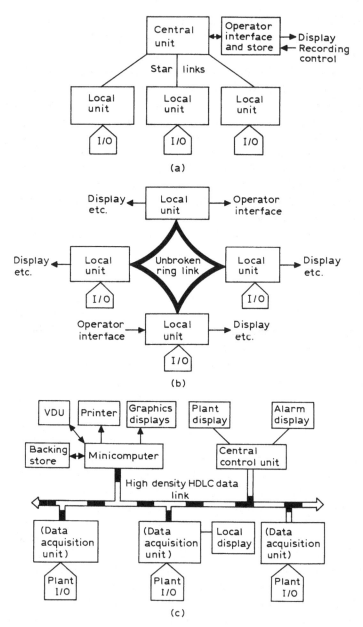

**Fig. 16.6.** (a) Star configuration; (b) ring configuration; (c) multidrop configuration.

Both the 'ring' and the 'multidrop' data highways shown in Fig. 16.6 provide a solution to this problem. Messages can be passed from any sub-system to any other and there is no need to have a 'master' unit at all. In practice, a master is often designated for 'traffic control' purposes: often the 'master' unit can be any of the sub-systems, and in some systems control can be passed from one to another unit. The ring topology has the advantage that messages can be passed in either direction around the ring: as a result one break in the communication link itself can often be tolerated, so long as the 'traffic' is not too close to saturation rate. The multidrop topology on the other hand can be extended more readily. The data highway communication system uses the principle of addressing, which is the basis of bus communication between processor, memory and other devices. Each sub-system must construct message packages which comprise several 'fields' of data: these include address, control, error and data itself. Because the communication link serves a common purpose, it has to be shared, and time division multiplexing (TDM) techniques are used to achieve this sharing. The functions of addressing, control and also of error detection require a processor, and the message packaging and control of data transfer for each sub-system is usually separated (with buffer memory) from the other functions of the sub-system.

The star and data highway topologies are sometimes combined in instrumentation systems. A local data highway is used to interconnect a number of local units such as data-acquisition units or process controllers, whilst at the same time each communicates independently by a star communication system with a powerful central control unit which combines the functions of display, recording and process control, etc. This is sometimes referred to as 'split architecture'.

The way in which any data highway is shared between the sub-system units 'sitting on it' depends on the functions of the system. It is rarely if ever the case that each sub-system will require equal and regular time intervals for data transfer. For this reason simple TDM multiplexing is never used. Each unit must be able to use the highway whenever it wishes within certain time limits for a limited period. Hence one fundamental constraint is the allowable length of a 'message' frame.

Whenever one unit is using the highway, all the other units must be aware that they cannot do so at the same time. Thus at least part of every 'message' may be received and interpreted by every unit in the system — the start and end 'flags'. Given this rule a 'collision' strategy can be adopted whereby, should two units happen to exactly coincide in transmitting a 'start' flag, both will give way. Since such coincidence in timing is most unlikely, not only will this rarely occur, but also when the two units try again it is virtually certain a second coincidence will not occur. This strategy works well in a system which does not have strict timing constraints such as may be imposed by instrument signal sampling techniques (see Section 15.1). A solution to this difficulty may be to send timing information about the sampled data within the message to be used when the analog signal is reconstructed at the receiving end.

An alternative to collision strategy is to have one unit in the system organise 'traffic' on the data highway. This organiser will of course have to service requests to use the highway and at times make decisions as to priority. This may be necessary in a system which combines alarm and shut-down data with less urgent instrument signalling. A version of this strategy is known as 'token passing', which involves passing a 'token' giving the right to use the bus from unit to unit (perhaps in strict sequence). The unit having the 'token' can transmit a single frame and then pass the token on to the next unit, for instance. Data transmitted may be addressed to a single receiving unit or alternatively to a group of units or even to all units, as required.

The decision as to what strategy is adopted as the basis for the 'bus protocol' is fundamental to the system design.

# CHAPTER 17

# *Communication Interfacing*

## 17.1 INTRODUCTION

Sub-systems or units of a distributed system will have to send and accept 'messages' over a shared data highway. Such messages will have to be addressed in some way to the unit concerned. In the simplest cases two units will communicate over a single 'dedicated' pair of conductors. However the communication link comprises receiving and transmitting equipment as well as the actual connecting cable, and for this reason it is necessary to design any communication link to a standard. Since in some cases receiving and transmitting equipment may not have been manufactured by the same company, the standards used cannot be simple company standards. The interface standard then must define:

1. physical details of plugs and sockets or other types of connecter to be used, the number of pins and the allocation of signal lines to pins;
2. levels of voltage or current flows to be used for signalling and tolerances allowed;
3. details of design of the receiving and transmitting equipment systems (including speed of operation, method of signal coding (FSK or simple voltage change), etc.);
4. 'protocol' for the transfer of data.

Protocol defines the rules for transfer of data between sub-systems over the system communication links or data highways. Several 'levels' of protocol may be needed, as will be seen later, the lowest comprising the 'handshaking' procedures by which data is transferred between sub-systems which may operate at different speeds and under other different constraints. Higher levels are concerned with the structure of messages

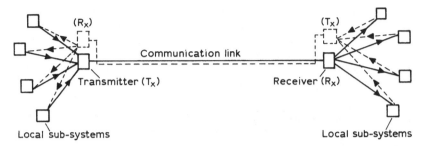

**Fig. 17.1.**

and the way in which a data highway can be shared between a number of 'users'.

In some cases the communication link may be of considerable length and therefore too expensive to be dedicated to the communication between a single pair of sub-systems (see Fig. 17.1). In such cases the link will be shared between several sub-systems at each end. The transmitting/receiving equipment for such links will have to be more complex than for a simple dedicated short-distance link, since it will have to deal with addressing of messages, multiplexing by TDM or FDM methods and frequency shift keying FSK coding techniques.

In other cases sub-systems, distributed over a wide area, may have to communicate with each other over a shared 'highway' (see Fig. 17.2). The transmitting/receiving equipment, whilst it will probably not have to deal with FSK modulation/demodulation of signals, will have to provide for the organisation of 'traffic' on the 'highway'.


## 17.2 DESIGN OF TRANSMITTING AND RECEIVING EQUIPMENT

The transmitting equipment and the receiving equipment may well be designed and made by different manufacturers, though usually this is not the case. The circuit which is set up via the interconnecting cable must therefore be defined by the interface standard. *The way in which this circuit is defined determines how susceptible or otherwise the data is to interference by the mechanisms described earlier.* For the relatively short distances which are usually involved in distributed instrumentation systems (a maximum of 2 km), voltage level changing (as opposed to frequency shift keying) is used to generate a signal. In its simplest form

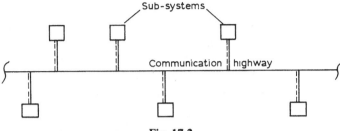

**Fig. 17.2.**

zero volts is maintained over one pulse time interval to represent a '0' state in the data, whilst the maximum signalling voltage used is maintained over a pulse interval to represent the '1' state.

In order that a circuit fault cannot be mistaken for the '0' state, 'bipolar' signalling, which uses a positive voltage to represent the '1' state and an equal negative voltage for the '0' state, is often used (Fig. 17.3). Thus the circuit shown in Fig. 17.3 can be set up between transmitter and receiver: the signal voltage is developed between the conductor ab and the conductor cd which is strapped to ground at both ends of the circuit. If ground potential (voltage) is different at the transmitter and receiver due to earth leakage currents from heavy electrical machinery, an alternating voltage will be induced in the circuit cdef which will add into the signal and cause interference. Such a circuit is said to be 'single ended'. At the same time both conductors of the communication cable are subject to common-mode interference from electromagnetic and electrostatic causes. Since conductor cd is grounded whilst conductor ab is not, the common-mode potentials induced are likely to be different in each and will not then cancel each other out but will again add into the signal. Clearly such an interface can only be used between locations where ground potential is known not to be affected by leakage currents from electrical machinery and over routes which can be protected from,

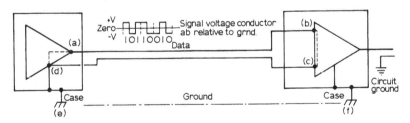

**Fig. 17.3.**

or are known to be free from, electromagnetic and electrostatic influences of a magnitude likely to cause interference. The simplest protection against all these causes of interference is to use a relatively large signal voltage so that the interference is of no consequence, but this is not always desirable (for electrical safety reasons for instance).

The circuit (between transmitter and receiver) shown in Fig. 17.4 incorporates several design features which improve immunity from all these forms of signal interference immeasurably. Some or all of these features are incorporated into the better interface standards. Equipment designed to implement the provisions of such standards is much more expensive than equipment designed to simpler standards, but is often required in industrial instrumentation systems.

Two signals are generated by the transmitting equipment in Fig. 17.4, one of which is the negative version or mirror image of the other as shown. Using a differential amplifier (Fig. 17.5) in the receiving

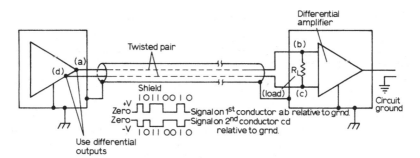

**Fig. 17.4.**

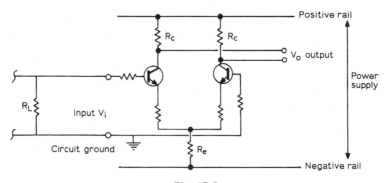

**Fig. 17.5.**

equipment, the difference in voltage between these two signals can be detected without any part of the transmission line circuit being strapped to ground as is the unavoidable situation with the simpler standard. Common-mode interference will affect both signals equally and will cancel out completely, so that even if the common-mode interference voltage is much greater than the signal it will not cause interference. Electrostatic interference is screened from the signal conductors by an outer braided metallic 'screen' which is placed around the insulated signal conductors before the outer insulating sheath is applied during manufacture of the cable. The 'screen' must be connected to ground at only one end in order that any current induced in it by electrostatic or inductive mechanisms will 'drain' away to ground. Further protection against electromagnetic interference is provided by twisting together the two insulated signal conductors: electromagnetically induced voltages in adjacent sections of the twisted cable will have opposed polarities as a result of this twisting, and will therefore cancel each other out to a substantial extent.

The input signal conductors are connected to the bases of the two matched transistors in the differential amplifier, as shown in Fig. 17.5. The sum of the currents flowing through both transistors flows through resistance $R_e$ and will vary slightly if the ground potential of the signal conductors and the 'negative rail' potential of the amplifier vary. This is not important, however, for provided the currents in both transistors are equal, there will be no difference between the voltages developed across both resistors $R_c$, and therefore zero voltage output $V_o$. If, however, a difference in voltage appears on the signal lines, the currents flowing in the two transistors will be different and therefore the voltage drop across the two resistors $R_c$ will be different, providing a voltage difference at the output. This voltage difference does not depend in any way upon the voltage level (with reference to 'ground') of the input (signal) circuit but only on the *difference* in voltage between the two conductors. The signal cables and transmitter, which are strapped to *transmitter* ground may therefore 'float' relative to *receiver* ground potential without affecting the signal in any way. In this way a ground return loop is avoided.

## 17.3 CABLE CHARACTERISTICS

Unit length of cable will have a capacitance C and inductance L; values of which are dependent upon the cross-sectional area of, and separation between, the conductors. Energy is stored in the cable by charging the capacitance and by overcoming electrical 'inertia' caused by inductance. The 'impedance' offered to the flow of electrical energy through the cable reduces with the rate of change of voltage in the case of C, but increases in the case of L. Thus

$$Z_{cap} = k \cdot 1/C \quad \text{and} \quad Z_{ind} = (1/k) \cdot L$$

Since the 'impedance' (the alternating current equivalent of DC resistance) represented by both $C$ and $L$ is given by the ratio of voltage to current flow, $V/I$, it follows that both impedances must be equal, since voltage and current flow are common. Hence

$$Z^2 = L/C \quad \text{and} \quad Z = [L/C]^{1/2}$$

Thus the amount of (signal) energy which can be stored in a *unit length* of cable is determined by the 'characteristic impedance' $Z_c$ of the cable, which is given by $(L/C)^{1/2}$. It can be shown that the signal velocity is independent of the characteristic impedance and of the transmitter/receiver. It follows that if the cable were of infinite length (endless) the rate at which energy could enter it and be stored would be in proportion to the cable characteristic impedance. If then the cable is terminated at the receiver in a resistance equal to the cable characteristic impedance, it will appear *to the advancing signal* to be of infinite length and the energy will flow into the termination (the receiver) at the same rate that it is flowing through the cable: thus all the energy will be developed as signal power ($I^2 Z_c$) in the receiver. If the cable is terminated in a larger or smaller value of impedance, however, the power requirements of the receiver will not be 'matched' to the line characteristics. Under these circumstances some of the signal power is reflected back along the cable, just as waves reaching the sea shore are reflected back if the angle of the beach (the terminating impedance) is not ideal. Just as waves reflected by the beach dissipate their remaining energy in collisions with advancing waves, so too signal waves reflected back waste some of their signal energy and reduce the potential data transfer rate. The cable should always terminate in a resistance equal to the cable characteristic impedance. The 'load' $R_L$ (Fig. 17.4) illustrates this. However, it is not always convenient, or even possible, to meet this requirement in simple equipment systems used for relatively short distances. Larger than necessary signal voltages are therefore used to compensate for the 'line losses' inevitable using such equipment.

The speed of data transfer possible depends upon the ratio of signal energy/unit time to the 'noise' energy and thus on the ratio $L/C$. It also depends upon the degree of pulse 'degradation' which can be accepted at the receiving equipment and hence on the level of interference (signal-to-noise ratio) which must be tolerated. Cable characteristics can only be improved at a price; coaxial cable may be used instead of a simple twisted pair. Very high data rates are necessary over a data highway used in a distributed instrumentation system over comparatively short distances. The cost of the cable is therefore of little importance for such a system, unlike a long-distance SCADA system. The data rate over such highways may be measured in millions of bits per second (MBaud) if the highway is protected from industrial interference: much lower data rates must be tolerated, however, where this is not the case (for instance, for highways connecting 'field' instrumentation or local control rooms). It is worth mentioning that in such cases optic fibre communication links may offer considerable advantages, since they are totally immune to interference of this type!

## 17.4 'SCRAMBLING'

For any given communication link, the rate at which signal energy can enter the line (signal power) from the transmitter is limited by the characteristic impedance of the cable (Section 17.3). Since energy flow rate into the line, which depends upon the rate of change of voltage (i.e. signal), is different for a '1' state and a '0' state, it follows that the power varies with the pattern of the 'bits' in any particular message. The voltage levels (and hence signal power) must therefore be limited to the worst case and the full potential of the link is thus not realisable. This reduces the maximum data rate possible. A technique which is used to even out the energy flow rate is to recode the message by rearranging the 'bit pattern' at the transmitter to give a more constant 'power spectrum', thus increasing the energy capacity of the link and hence the data rate.

## 17.5 SYNCHRONISATION

The transmitter real-time 'clock' sets the duration of each 'bit' ('0' or '1') and it is essential that the receiver should allow *precisely* the same interval of time for each and every 'bit'. If it does not, a cumulative error in the position of the start time of each bit will build up, and eventually the receiver may confuse two successive bits. Over long distances 'line delays' tend to vary significantly, making such precise matching of transmitter and receiver clocks even more difficult.

Provided messages are not too long, transmitter and receiver clocks are accurately synchronised and variability of line delay is not significant, the cumulative error may be small enough to be tolerated. Thus 'asynchronous' operation is suited to short messages transmitted over short distances. A single 'start' bit at the beginning of the message or 'frame' is used by the receiver to synchronise its clock; thereafter the receiver must rely on its clock pulse intervals being exactly the same as those of the transmitter (Fig. 17.6(a)). For long lines or for fast bit rates which, by definition, mean a short pulse duration, it is necessary to send 'timing information' as a separate transmission. Operation is then 'synchronous' rather than 'asynchronous'. This usually necessitates a second 'timing data' communication channel within the communication link, thus reducing the capacity of the link in FDM multiplexed operation, or necessitating a second pair of conductors where solid wire communication channels are used. Much higher data rates are possible, since synchronisation occurs at the beginning of every 'bit' interval instead of only at the beginning of each message. Alternatively, alternate 'frames' of eight bits can be devoted to timing information rather than data, as shown in Fig. 17.6(b). This TDM multiplexing reduces the data transfer rate by half and it is once again debatable whether this offers any advantage over the use of a separate channel. An additional 'timing channel' can also be avoided by using 'Manchester' coding in which the voltage level changes half way through each 'bit' (Fig. 17.6(c)). Unlike simple voltage-change signal coding, timing information is provided in Manchester coding by this voltage change. However two rather than one voltage level changes occur during each 'bit' interval, and the capacitive/resistive characteristics of the cable will cause greater degradation of the signal and hence reduce the signal-to-noise ratio. In 'noisy' environments the data transfer rate will be reduced. On the other hand, Manchester coding uses constant signal power because '1's and '0's take equal power; thus 'scrambling' is unnecessary.


## 17.6 TRANSPARENCY

The concept of 'transparency' necessitates that a function be carried out without placing *any* constraints on the equipment at the input or output.

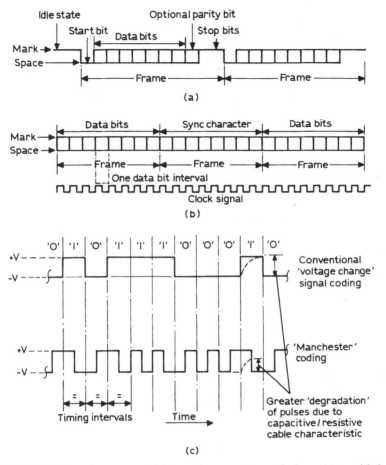

**Fig. 17.6.** (a) Asynchronous and (b) synchronous transmissions; (c) combining data and timing information.

Ideally, a communication channel should be 'transparent' to the transmitting and receiving sub-systems. This will necessitate that the communication system include both buffer memory and queuing facilities: multiplexing, data transfer speed changing, error detection and correction and other functions must be performed by the communication channel rather than the transmitter and receiver. If radio telephony links are used, 'modulation' and 'demodulation' functions, to encode the data as an FSK signal, will also have to be 'transparent' to the transmitter and receiver. The specification of any communication system is designed with this concept in mind.

## 17.7 LONG-DISTANCE COMMUNICATION

For solid-wire communication links, the interface between the two microprocessor units (or sub-systems) consists of the transmitting equipment, the cable, line drivers if necessary, and the receiving equipment. If it is necessary to communicate over a long distance, frequency shift keying techniques must be adopted in order to reduce the pulse distortion (see Section 16.4). A separate piece of equipment must be installed after the transmitter and before the receiver at each end of the communication link: this is the 'modem' or modulator/demodulator. The transmitting modem generates one of two different sinusoidally varying alternating voltages when it receives a '1' or a '0' from the transmitting equipment (modulation), whilst the receiving modem demodulates the different frequency signals to generate the original voltage level change signals in front of the receiving equipment. Where the communication link is a solid conductor cable, 'limited distance modems' using FSK techniques can extend the communication distance at 20 kbaud without line drivers from about 1200 m to about 15 km. Above this distance the communication link itself will be a telephony or radio transmitter/receiver channel using FSK techniques within a defined 'bandwidth' which is only a small fraction of the full bandwidth of the channel. Using FDM multiplexing, the full bandwidth of the channel can be shared between a number of transmitting/receiving pairs of sub-systems. However, the process of modulation and demodulation required to 'compress' the data into a very limited bandwidth is very much more complicated than in the case of limited distance modems which use the entire bandwidth of a solid conductor cable. The techniques of radio telephony signalling are beyond the scope of this book.

In addition to its main function of signal modulation/demodulation, a 'modem' carries out other functions associated with transparency:

1. Error detection is a function of the modem. This includes organisation of data and addition of appropriate 'error bits' at the transmission end, counting of bits and error detection at the receiving end. Forward error control or FEC (see Section 6.6) is used to correct errors if possible without requesting a repeat. Together with ARQ or automatic request to repeat, this technique usually provides adequate speed of data transfer. Under bad interference conditions the error rate may be so high that the

message repeat rate results in the effective data transfer rate being seriously reduced. Under such conditions the modem may automatically revert to half-speed operation. This means that either the frequency at which messages can be transmitted by any sub-system over the communication link, or alternatively the length or number of the messages, must be reduced. For an instrumentation system this may mean that the sub-systems must reorganise the data packaging to reduce the number of measurements being transmitted, leaving out non-essential measurements temporarily, for instance. After an interval of time the modem will attempt to revert to full speed operation again, but will again reduce to half speed if the error rate is still excessive.

2. Transmitter/receiver synchronisation is a function of the modem. A choice of synchronous or asynchronous operation is usually provided.

3. The process of multiplexing the signals is a function of the modem in radio telephony communication systems.

4. 'Elastic buffer memory' is often provided within a modem to permit it to store data messages for transmission or on reception temporarily. This enables the communication system to operate independently of the sub-systems to a large degree (transparency). It also enables use of the communication channel to be maximised.

5. Some modems use 'scrambling' techniques to maximise the data transfer rate.

6. Any communication channel will have to have both a transmitter and a receiver at each end. However, the particular requirements of the system may make it necessary for data to be transmitted in both directions simultaneously or only in one direction at any time. In the former case it is obviously necessary to provide two pairs of conductors (or two radio channels) and separate receive and transmit modem functions at each end. It is then said that there are two 'half-duplex' channels. If on the other hand transmission of data in one direction or the other at any time is adequate, then a single 'full-duplex' channel will be used. The modem at each end must still contain a full set of both transmit and receive functions but, in addition, it must be able to implement handshaking protocol for controlling the direction of data transfer. (Since the cost of a half-duplex channel is more or less the same as the cost of a full-duplex, there is obviously no

point in providing two half-duplex channels unless the data transfer rate in both directions is uniformly equal, or to meet some special requirement.)

## 17.8 INTERFACE STANDARDS

One of the most widely used interface standards is the American Electronics Industry Association (EIA) standard RS 232c, which covers 'bit-serial' data transfer. It defines all necessary 'handshaking' control signals required between the serial communication interface of the microprocessor sub-system and the communication link or (if a solid-wire link) the receiving sub-system, over distances up to 15 m at data transfer rates up to 20 kbauds. It specifies a single-ended bi-polar voltage unterminated signal circuit.

A nominal $+12$ V signal is used to represent binary state '1' and $-12$ V to represent binary state '0'. This removes the possibility that an open circuit or 'lost' digit will be construed as a '0'. Tolerances are very wide, any voltage between 3 and 25 being 'recognised', which means that circuit components constructed to this standard must be capable of withstanding voltages between $-25$ and $+25$ V in operation. The protocol defined by RS 232c necessitates 25 signal lines and a 25-pin plug-and-socket connector. A significant number of these lines are rarely needed, as RS 232c, like any standard, provides for more functions than are normally required. The signal lines and pin allocation are shown in Fig. 17.7.

The first 15 signal lines are used to control actual data transfer, regardless of whether solid wire or telephone communication channels are used. The remaining 10 lines are devoted to control of a modem when the RS 232c link connects a sub-system to a communication system for data transmission over long distances. How many of the 15 data control lines are used for a particular application depends on the design of the equipment; for this reason it cannot be assumed that two units from different manufacturers will necessarily work together, even though they are both said to conform to the RS 232c standard. This apparently lax approach to standardisation is foreign to engineers and may seem unfortunate, but it must be seen against the background of very rapid changes in this industry.

There are two major limitations with the simple RS 232c standard. First, it provides no implicit means to transfer timing information.

## RS-232C LINE DESIGNATION

| Line pin no. | Circuit name | Description |
|---|---|---|
| 1 | AA | Protective ground |
| 2 | BA | Transmitted data |
| 3 | BB | Received data |
| 4 | CA | Request to send |
| 5 | CB | Clear to send |
| 6 | CC | Data set ready |
| 7 | AB | Signal ground (common return) |
| 8 | CF | Received line signal detector (data carrier detect) |
| 9 | — | (Reserved for data set testing) |
| 10 | — | (Reserved for data set testing) |
| 11 | | (Unassigned) |
| 12 | SCF | Sec. rec'd. line sig. detector |
| 13 | SCB | Sec. clear to send |
| 14 | SBA | Secondary transmitted data |
| 15 | DB | Transmission signal element timing (DCE source) |
| 16 | SBB | Secondary received data |
| 17 | DD | Receiver signal element timing (DOE source) |
| 18 | | Unassigned |
| 19 | SCA | Secondary request to send |
| 20 | CD | Data terminal ready |
| 21 | CG | Signal quality detector |
| 22 | CE | Ring indicator |
| 23 | CH/CI | Data signal rate selector (DTE/DCE source) |
| 24 | DA | Transmit signal element timing (DTE source) |
| 25 | | Unassigned |

**Fig. 17.7.**

Consequently data transfer must be 'asynchronous', and hence slow. Secondly, the voltage levels in a 'single-ended' circuit are strapped to ground potential separately at both ends of the cable. The 'ground loop' implicit in this ground return circuit introduces 'noise' into the signal circuit in industrial environments or steel-framed buildings with rotary electrical machinery. Because of these limitations the standard is inadequate for data transfer over more than a few metres at moderate speeds, and totally unsuitable for an industrial environment.

RS 232c is often used to transfer data over a short solid-wire link to a modem, which controls transfer of data over a telephone or radio communication channel. In this case some of the other 10 modem control lines will be used in addition to those required to implement the protocol of actual data transfer between the two sub-systems.

CCITT counterparts of EIA RS 232c are Recommendations V.24 and V.28, which are essentially the same but differ in the signal pin nomenclature. The military standard equivalent of RS 232c is MIL 188c (low level).

RS 422A is a standard which specifies a differential balanced-voltage signal circuit, terminated in an impedance equal to the cable characteristic impedance (about 100 $\Omega$ for a typical 24 AWG twisted-pair cable. CCITT counterparts are V.11 and X.27 and MIL 188-144 (balanced)). It is capable of a 100 kbaud data transfer rate over distances of up to 1200 m (RS 232c — 20 kbaud over 15 m) and can withstand interference causing unwanted 'noise' voltages up to $\pm 1$ V (measured by substituting a 50 $\Omega$ resistor for the receiver). Moreover, because it does not specify a 'single-ended' circuit but requires instead a single ground connection for a two-wire signal circuit, which necessitates use of differential amplifiers in the receiving equipment, it has a high degree of immunity to ground-loop and common-mode interference.

It is not possible to mix RS 422A and RS 232c receivers/transmitters because the signal circuits specified by the two standards (one single-ended and the other balanced-voltage) are incompatible.

Like RS 232c, RS 423A is a standard that specifies a single-ended, bi-polar, unterminated signal circuit. However, unlike RS 232c it specifies differential amplifiers in the receiving equipment, and in consequence can operate at up to 3 kbaud data transfer rates over 1200 m (compared to 100 kbaud for RS 422A) and 300 kbaud over short distances of about 15 m (compared to 20 kbaud for RS 232c). It is possible to use RS 423A receivers with RS 232c transmitting equipment (which includes a large range of equipment on the market) because both standards specify single-ended bi-polar signal circuits.

CCITT equivalents are V.10 and X.26, military equivalent MIL 188-144 (unbalanced).

# CHAPTER 18

# Message Packaging and Protocol

## 18.1 INTRODUCTION

'Protocol' is the collective name given to the rules which are laid down to govern the interaction of the sub-systems within a total system environment. In system terminology the total system is always 'closed'. That is to say, the total system is absolutely self-contained and is not influenced in any way by conditions outside. In practice, however, this is not a very useful concept as it is rarely fulfilled. Nevertheless it is a concept which helps to define the 'highest level' of protocol required: this is the protocol required to govern interaction between the largest sub-systems within a total system. Putting this into the context of the international telephone system, there are obviously a number of levels of protocol required, starting at the lowest level, with the protocol which governs each individual subscriber's interaction with the local exchange, and finishing at the highest level with that governing the interactions of national 'networks' which are sub-systems of the international system. In the context of industrial instrumentation systems, however, the 'total' system may be bounded within the plant, or may possibly include a group of plants, in which case the highest 'level' of communication interfacing will take place over telephony equipment systems.

The lowest level of protocol required in an instrumentation system defines the handshaking requirements between individual micro-processor-based units over single-pair dedicated communication links (for instance, between a single process controller or instrument and a data-gathering sub-system). These have already been described in Section 14.4. The communication between data-gathering systems and other major sub-systems of the total distributed instrumentation system over a shared data highway necessitate a different and more complex set of rules and can be regarded as a second level of protocol. The rules which constitute second-level protocol control the following functions:

1. To establish and terminate connections between two sub-system units connected to the common highway.
2. To ensure message integrity through error detection, request for retransmission and control of the speed of data transfer.
3. To identify the sender and receiver of messages.

One of the functions of second-level protocol is to provide 'data transparency', which is to say that the control data, error bits, etc., must be treated separately from the message data by all equipment using the protocol.

## 18.2 SECOND-LEVEL PROTOCOL

The essence of second-level protocol is that data messages or 'frames, as they are sometimes referred to, are transmitted over a TDM shared communication link. Each 'frame contains several designated 'fields' of data, only one of which is allowed to contain the message information. All the other fields are associated with the protocol functions. The construction of a typical frame is shown diagrammatically in Fig. 18.1.

The 'flag' field contains a number of bits (say 8) which together indicate to a potential receiver that a message is about to be sent. The same flag sequence indicates the end of the message. Thus whenever the receiving circuits in any sub-system decodes this particular and unique sequence of bits it is alerted that a message is beginning or ending. The receiver of every sub-system connected to the communication link or highway receives *all* data which is put onto the communication link; when it detects a flag sequence it decodes the next field which contains the address of the sub-unit which is the intended recipient of the message. If the address matches its own it will 'accept' the data which follows into its own buffer storage memory until it detects a second flag

| 'Flag' field 8 bits | Address field 8 bits | Control field 8 bits | Information field  Variable length | Error check field 16 bits | 'Flag' field 8 bits |
|---|---|---|---|---|---|

**Fig. 18.1.** BOP (SDLC) message frame format.

sequence denoting the end of the message. If the address does not match its own it will ignore the data following until it detects the next flag sequence (end of message) which tells it that the next flag sequence may again be a message addressed to it. There is an obvious risk that the flag sequence will by pure chance occur within the information field. Since the receiver will take this as the end of the message this must be prevented. This is achieved by a technique called 'bit stuffing': the transmitter scans all data before transmission and if it discovers the flag sequence it adds into it an additional bit pattern which breaks up the flag sequence within the information field. After the message has been accepted into the receiving system buffer memory, it is scanned by the receiver, which discovers this special 'stuffed' sequence and removes it before decoding the message. Lastly, the receiver will carry out error detection using the error detection field bits and will if necessary request a retransmission. If however there is no error or the error can be corrected without retransmission, then the message itself, comprising the control and information fields of the frame, are made available to the sub-system in a buffer storage memory. Thus the communication system is 'transparent' to the sub-system, which need know nothing of the functions defined as protocol.

The message thus transferred over the communication link from one sub-system unit to another comprises a control field which tells the recipient sub-system what it is required to do. For instance it might be a data acquisition system, in which case it might be instructed to scan certain instrument transducers and transmit measured values to another sub-system. It might be a process controller, and be instructed to change its control actions to new values which would then be contained within the information field which follows.

## 18.3 THIRD-LEVEL PROTOCOL

Nothing in the foregoing description of second level protocol functions determines the way in which 'traffic' on a multi-user data highway is controlled. Such functions are considered to be a third level of protocol even though it may be difficult to separate out the hardware and software which implement them. As mentioned in Section 17.1, a number of protocol philosophies are available, depending on the functions of the total system served by the communication highway. A SCADA system comprising a central display, operator interface and

historical data recording system, together with a number of widely
separated data acquisition units, will probably utilise a star configuration
of dedicated communication links. In such a system the third-level
protocol would be based on central control of traffic, since the only
route from any satellite data acquisition unit to another would be via the
central facility. The protocol would in fact be exactly the same as the
interrupt system of a processor, with priorities established as necessary.
The protocol of a distributed system using a common data highway
would be quite different and would be based on a philosophy of equal
priority with 'collision' protection in all probability. The design of this
protocol and the hardware or software to implement it is implicit in the
design of the system itself. One of the fundamental factors in the design
of such a system, the communication highway and the third level
protocol, is the estimation of the data rate required. It is all too easy to
underestimate the required data rate.

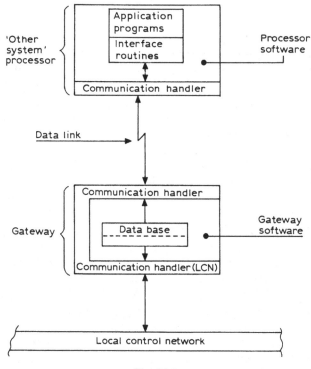

**Fig. 18.2.**

For example:

A system comprises 10 data acquisition units, each of which collects measured value data from 10 analog transducers. Each measured value is sampled 10 times per second and a resolution of $\pm 0.1\%$ is required of the digital coding of the analog signals.

For a resolution of 1 in 1000 (0.1%) each analog value must be represented by $n$ bits where $2^n$ is greater than 1000. Hence $n = 10$ bits ($2^{10} = 1024$). A further 4 bits are required in the message construction to identify each of the 10 measurements taken by a single-data acquisition system ($2^4 = 16 > 10$, $2^3 = 8 < 10$). Thus the information field requires at least 140 bits. Adding 16 for the two flag fields, 8 for the address field, 16 for error detection, and 8 for the control field, and the minimum frame length is 188 bits. Thus 10 frames per second for each of 10 data-acquisition units requires that the data highway operates at a data transfer rate of 18.8 kbaud. This is a comparatively small system and there will obviously be a need for additional information space in each message frame. If the data highway is about 1200 m in total length, then the data transfer rate is well beyond the capability of either RS 232c or RS 423A interface standards without the use of line drivers.

## 18.4 COMPATIBILITY OF PROTOCOLS

The International Standards Organisation (ISO) defines seven levels of protocol to deal with the transfer of digital data. This 'open system model' as it is called, is important to telecommunication engineers but only the lowest levels concern the control engineer. The concept of the 'gateway' to connect two systems, produced by different manufacturers, is a product of the ISO initiative. The gateway itself (see Fig. 18.2) can be produced by either manufacturer.

The protocol of the gateway must be available to other manufacturers and it must incorporate a special type of data base for data to be transferred in either direction. One section of this data base will be 'written to' only by system 'A' and 'read' only by system 'B', whilst the other section will be 'read' only by system 'A' and 'written to' only by system 'B'. The protocol must take care of the collision problem. Messages between either processor and the gateway are 'queued' and organised by a 'communication handler' program. The concept of the gateway allows equipment manufacturers to design a piece of equipment (the gateway) which will enable data to be transferred between system highways which are themselves operating at different speeds and under different bus traffic management protocols. It is a clumsy expedient, but it has to be borne in mind that each manufacturer is (rightly) more concerned, in the medium term, to ensure compatibility between new and older generations of the equipment systems he sells, than to make new equipment systems compatible with other manufacturers equipment.

Perhaps of more interest to process-control equipment manufacturers is the Manufacturing Applications Protocol or MAP which was initiated by General Motors but is now an internationally supported standards body. The concept behind MAP is of a plant-wide data highway which will form the 'backbone' of a total automation system in which even CAD (Computer-Aided Design) systems can communicate, if required, with local controllers or robots, or anything else for that matter. This 'backbone' highway is seen as carrying computer, television, facsimile, telex and speech as well as automation and instrumentation data channels. It will link together different systems and provide for total integration of all digital plant equipment from telephones to control systems eventually.

The plant 'backbone' envisaged by the MAP standards body will have to be a coaxial-cable broadband-frequency multiplexed highway in order to be capable of carrying all the 'traffic' envisaged. It is possible however that fibre optics may be used ultimately. It is quite common, even today, to find plants which have some form of 'backbone' highway (usually based on Ethernet or some other computer standard), though the majority of current process-control and building management systems use the gateway concept to interconnect different buses, even when these are produced by the same manufacturer!

Future systems will have to conform with the protocol defined by MAP or the ISO model, from intelligent transducers up to central control rooms, from telephones and facsimile machines to desk-top computers and control systems. In order to maintain their markets, equipment suppliers will be forced to conform to these protocols. This will be made more certain in time and at the same time easier to achieve, by the fact that manufacturers of VLSI chips will also be forced to standardise on circuitry to support standard protocols. Indeed they will be only too ready to do so since the volume of standard chips will be enormous and the price correspondingly low. Standardisation of protocols will have a snowball effect on the entire industry, which will have incalculable results for process control. However, because of the necessity for intra-company standardisation it may be a long time before such industry-wide standardisation comes about. Definition of details, such as the message format for measured value data and the protocol of a 'field bus', (by which is meant the data highway of the distributed control system) are likely to be finalised later than the 'higher' levels of protocol. How special requirements such as intrinsic safety (electrical/fire hazard) will be met by standardised VLSI chips is anybody's guess at the moment, but undoubtedly they will be eventually. Indeed, such is the pace of development of technology in these areas, and such is the pressure internationally for standardisation in this area, that it is likely that the process will be well under way by the time this book is published.

# PART 4

# CHAPTER 19

# *The Distributed Control System*

## 19.1 INTRODUCTION

The processing power which can be made available cheaply and exclusively in a single sub-system has made it possible to replace the inflexible and unwieldy multiplicity of indicators and recorders previously common, (even with the earlier DDC (direct digital control) computer systems) with a single computer VDU display and operator interface. The graphics capability of digital computing equipment and the use of interactive programming (see Section 13.6) have made it possible for multiple displays to be available, at the touch of a key, on a single display screen. Thus a hierarchy of 'overviews' and more detailed 'views' of the plant, generated by compiled programs written by the equipment manufacturer, are made available to the operator in the form of process flow/instrument diagram graphics. The set point, alarm and even control action data can be shown on these graphic displays and updated rapidly from the data obtained by other data-acquisition subsystems. Interactive programs supplied with the equipment enable set points, alarm values or control action data to be changed from a keyboard. The size of the control room has thus been greatly reduced, representing a major financial and logistical advantage by itself.

DDC introduced the advantages of accurate computation into large regulatory control systems comprising a number of SISO control loops and, in earlier analog systems, several 'computing relays'. Such computation, however, placed considerable additional load on the single, already overloaded, central processor. The fact that units can communicate quickly and directly with any other unit in the plant, using a common data highway, without tying up dedicated communication channels to a central processing unit, is an immense advantage for distributed systems. Interaction between controllers and other units situated at any point in the plant becomes practicable. Thus, much more powerful ratio, cascade and multivariable control strategies are capable of practical realisation than hitherto. This last advantage seems to have been largely neglected up to the present time, possibly because exploitation requires application of more advanced control theory than has hitherto been considered necessary in the process industries.

Direct digital control based on single (central) processor monolithic computers introduced the enormous advantage of 'historical data storage'. All the sampled data for many measurements over long periods could be stored accurately, something that was impossible with earlier, totally analog, instrumentation systems. Analysis of past plant operation, had previously involved laborious searching through paper chart records and extensive and manpower-intensive analytical work, and was in consequence rarely attempted. This now became extremely simple, providing an effective new management tool. Distributed processing has extended this facility to larger numbers of measurements and longer periods of history, which previously would have imposed an unacceptable additional processing load on the single central processor, by utilising an independent 'historical data' unit. Moreover, such historical data sub-systems can now carry out extensive analytical processing on the data, such as averaging and regression techniques. These developments in their turn introduced the possibility of on-line computer optimisation of the process operation, based on historical data analysis and computer 'modelling' of the process itself. Such 'process management' computers can themselves 'sit on the data highway' and so interact directly with, on the one hand, the data acquisition sub-systems, and on the other hand the set points, etc., of individual SISO regulatory controllers. Thus, open- or closed-loop 'supervisory' control is today quite practicable, which extends the scope of plant management still further (see Fig. 19.1).

Perhaps one of the most important advantages of the distributed instrumentation system is its ability to handle alarm and shut-down signals. Separation of the functions of data acquisition from those of logical processing of the data makes it possible to implement, in 'real time', the most complex of safety alarm and shut-down regimes. In practice, however, because of the different emphasis on functional security of safety systems, it is usually better to construct a totally separate 'safety and shut-down' system (see Chapter 22).

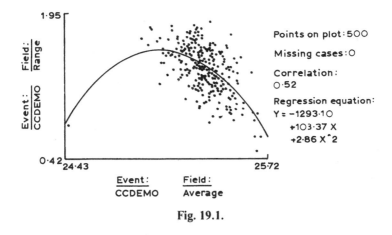

Fig. 19.1.

'Architecture' is the word used to describe the way the building blocks or sub-systems of a distributed processing system are put together to meet the specification of the system itself. Not only are the sub-units physically distributed around often very large plants, but also data-processing functions are distributed between multiple processors in these units. Data gathering and regulatory control functions can be decentralised without any disadvantages, since the control actions and set values can be 'downloaded' from a separate operator interface unit in the control room at any time. This decentralisation results in a very considerable saving in cable costs and this was undoubtedly the original incentive for equipment manufacturers to develop such systems. Such physical decentralisation also simplifies maintenance in comparison to earlier DDC systems, since it separates the hardware related to these functions. Incidentally it effectively removes the greatest disadvantage of the DDC system, that of commissioning the *entire* instrumentation system at a given stage in the plant-commissioning process, instead of piecemeal as was previously the case with analog systems and is again the case with distributed digital systems.

The data highway can be clearly seen as the key to design of any distributed system. It is possible to include in a single system any number of sub-systems, each carrying out a separate task and each utilising as many processors as necessary to carry out its own internal task/s. This proliferation of sub-systems within a single system correspondingly *increases* the 'traffic' on the data highway. Increasing distance between sub-systems, which often goes hand in hand with increasing numbers of sub-systems, *decreases* the data transfer rate possible. These two factors combine to make the architecture of the data highways the limiting factor in the design of any distributed instrumentation system.

## 19.2 THE DATA HIGHWAYS

Chapters 16, 17 & 18 dealt with the basic principles, interface standards and protocol of communication links in a general sense, as these apply to all types of distributed-processing microprocessor system, as well as long-distance data communication systems (SCADA). In this section therefore, the way in which the particular requirements of a distributed instrumentation system are constrained by these basic principles will be discussed only briefly. Each individual manufacturer in this field has by now introduced at least one distributed system to the market. However the associated technology is developing at such a rate that it should be no surprise to find that there are very wide differences in capability between available systems. To a considerable extent the capability will reflect the date that the system design was laid down and the commitment of the manufacturer to investing in further development and fundamental redesign. However the basic structure or architecture of all the available systems has much in common, since they all carry out the same functions, which are based on the very considerable advantages of distributed systems over all earlier equipment systems.

In the first place all systems will be found to distinguish between two fundamentally different types of data 'bus' or 'highway'. One of these will serve the purpose of interconnecting those sub-systems which, on the one hand carry out very large amounts of data processing and require communication at very high speeds, and on the other hand are located in the relatively interference-free environment of a central control room. The other type of data highway will be found to connect control and measurement devices, which are located in local control centres, into the high-speed data highway via communication interfaces. This later has to handle 'real-time data' and its protocol is therefore constrained by this fact. It is also, inevitably, routed through areas of high interference. It should be no surprise, therefore, to find that the second type of data highway operates at much lower data rates than the first, and that whereas the first type operates under a 'token passing' protocol, the second will operate under a rigid 'scan-time' protocol. Both will probably be of the multidrop type, as the ring architecture cannot readily be expanded, a feature which is obviously essential to any industrial instrumentation system. The first type of data highway may operate at anything up to 10 MBaud (10 million bits/s) data transfer rate, which is the limit set for balanced-voltage interface standards. The second type may typically operate at up to 250 kBaud which is a very high rate in view of the interference problem.

An equipment manufacturer can readily specify the levels of both electromagnetic and electrostatic interference that his equipment can tolerate, but it is almost impossible for a user to establish the maximum levels of either, which operation of his plant will be likely to impose. To do so would entail running every last bit of electrical machinery at maximum rating simultaneously — which would certainly give an unrealistically high result, making demands which the equipment could not meet — or establishing what the worst set of operating conditions would be — which is impossible to realise. Therefore only extended testing of actual equipment systems at the user's own plant can reliably establish whether the particular system can tolerate the interference which will be met either in the central control room or in the 'field'. At first sight it would appear to be a good practice to incorporate, as is common in modems used for telecommunication, automatic Baud rate reduction on error rate increase. It is not possible however to do this within a real-time system unless considerable redundant capacity is built in, which would make the equipment uncompetitive. Unfortunately this is a totally unresolvable problem, and the user will be well advised to ask the equipment manufacturer to accept the responsibility to ensure that the equipment will operate satisfactorily in the *actual* environment in which it will be installed. Manufacturers, however, are understandably wary of accepting contractural responsibility, as they have no control over the environment into which their equipment is to be installed. Nevertheless they must be prepared to accept this responsibility, provided they are given adequate opportunity to assess the environment, if they hope to market their equipment.

Since all data transfer is along a single cable, a distributed instrumentation system is vulnerable to accidental cable damage. Most equipment manufacturers supply dual communication highways which have the facility to detect loss of continuity in the 'active' cable, and automatically switch communications through the standby. Of course the two cables should be routed entirely separately for this facility to be an effective protection.

The type of cable to be used for these highways will be specified by the equipment manufacturer, and is vitally important to the satisfactory operation of the system. Again the user must purchase cable to the specification of the equipment manufacturer, who must also provide written details of any installation constraints so that these can be included in the installation contractor's conditions.

## 19.3 SYSTEM ARCHITECTURE

The typical architecture of the highways is shown in Fig. 19.2. Typically, a distributed instrumentation and control system has three 'levels'. The 'lowest' level, or that nearest to the plant itself, comprises the measuring devices, impulse transmission, transducers and signal transmission systems to local control rooms. Until recently all such local signal transmission systems have operated on established analog principles. It might have been expected that all transducers would by now generate digitally coded signals but, for the most part, this is not the case. There are two good reasons for this. First the original spur to development of distributed instrument systems — massive reduction in cabling costs — does not apply at this level. Each individual transducer must be connected directly to the local data-gathering point in a local control room; there would be little saving in a common ring-main data highway at this level. Secondly, and probably of more significance, are the constraints placed on the equipment system designer by the requirements, in many industries, to ensure that electrical apparatus does not present a risk of explosion or fire (see Section 7.10). A complex technology has built up around low-energy analog signalling systems since the late 1950s and the equivalent technology for digital equipment has been slow to develop to the same state of standardisation and certification. Until it does, however, this lowest level of the system architecture will remain predominantly analog. There are advantages to be gained by total 'digitalisation'; these are mainly in the area of remote calibration and fault diagnostics (see Section 20.1).

The second level of system architecture comprises the data-gathering systems in local control rooms, the slower real-time data highways and

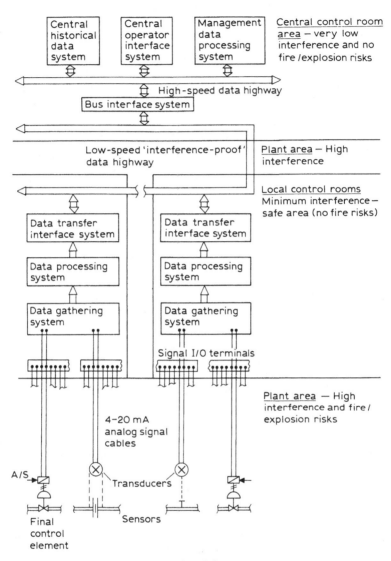

**Fig. 19.2.**

the communication interfaces with the faster data highways. Local control rooms must be 'safe areas' where fire/explosion risks exist. In comparison with earlier centralised DDC systems the cost of these local control rooms must be set against any capital cost savings on cabling.

The functions of measurement data gathering and of continuous regulatory control of process variables are vested in these equipment systems. In addition, the functions of packaging this data and controlling its transfer to a central operator interface unit ('sitting' on the fast data highway) and/or to other units anywhere within the system, are vested in this equipment. Data entering this equipment will comprise mainly 'downloads' of set-point data to controllers, requests for specific data which is not regularly scanned and diagnostics. Such set-point downloads may originate in process management/supervisory control computers or from the central operator interface. In addition, some extensive interactions in the form of ratio control loops, cascade control loops, etc., may involve regular data transfer between local control rooms at this second level without reference to the central operator interface or any other higher-level system. Such interaction with a multiplicity of other systems at both this second level and higher levels in the hierarchy are the reason why 'star' architecture has been largely abandoned in favour of the multidrop architecture. The actual structure of multiloop control systems (cascade, etc.) can be altered from the central operator interface (level 3) by downloading address data, so that the set point of a controller is taken as a different measurement value than before, for instance. In this respect the system is no different from earlier centralised DDC systems. If the measurement value to be taken as a set point signal originates in a different local control room, the message packaging and transfers between highways may be implicated in such a change; a special unit may be required, sitting on the faster highway, to handle such matters.

The third and final level in the architecture of a distributed instrumentation and control system comprises the faster data highway and all processing systems 'sitting' on it. It is usually located entirely in the central control room for reasons of fire and explosion safety and also to isolate it from electrical interference of all types. This latter reason is easily forgotten when powerful short-wave radio communication transmitters are installed in close proximity to the instrument equipment (sometimes in the same cabinets!). Even worse, the use of thyristor lighting dimmers in the control room can play havoc with both radio and instrumentation equipment! Again the manufacturer should be asked to supply written details of all such constraints on the operation of this level equipment, *in time for them to be included in contracts for other work.*

This third-level equipment will typically include one or more VDU and keyboard operator interface system which will incorporate all the manufacturer's special interactive programs to enable the operator to call up graphic displays, change set points and control structures and respond to alarms, etc. The equipment will also contain, in its memory, the compiled programs which actually generate the graphic displays on which 'live' real time data such as measured values are displayed and are 'called' by the interactive programs run by the operator. Also at this level there will almost certainly be one or more historical data storage unit, into which 'past' values of measured data are stored, and which will contain the compiled programs which enable graphical displays of a historic nature to be generated on the operator interface VDU.

## 19.4 THE SUB-SYSTEMS

The sub-systems of a distributed processing instrumentation and control system fall into two categories, according to whether they 'sit' on one of the slower data highways or on the faster data highway. The former type are designed to gather measurement data and carry out more or less fixed real-time processing, such as the calculation of PID control actions, on this data. Constants such as proportional, integral or derivative control action factors may be varied from time to time and set points changed by 'download' from one of the higher-level sub-systems such as the operator interface or a supervisory computer. Such downloaded data change variables in compiled and assembled programs executed by the processors in these units, which continue to process the measurement data at constant (real-time) intervals using the latest such variables supplied to them. The latter type of sub-system, on the other hand, is designed to carry out the more extensive processing operations on the data gathered, such as the generation and updating of graphic displays of 'live' or historical measurement data. These functions are not constrained to run in real time, but do involve the transfer of large amounts of data between such units as a historical data unit and the operator interface unit.

## 19.5 DATA-GATHERING UNITS

There are three types of data which have to be captured in real time from the plant transducers:

(i)   continuous analog measurements (usually 4–20 mA);
(ii)  pulse-rate signals representing integrated or 'totalised' flows, as opposed to instantaneous flow-rate analog signals; the pulses have to be counted;
(iii) discrete signals, generated by switching action, representing alarm or other such information.

To this list might be added digitally coded analog-sampled data signals.

In respect of analog signals, the data-gathering unit will need to provide the power supply for the individual 4–20 mA transducer circuits as well as multiplexing, sampling and A/D conversion. Each pulse-rate signal requires a counter which is then sampled at intervals and the value converted into binary form for data transfer. Discrete signal inputs can be read directly by the processor as a '1' (closed contact) or a '0' (open contact) and only require a single bit to represent each datum.

Sub-systems having a limited number of analog and/or pulse-rate signal inputs and a larger number of discrete signal inputs are often provided, in order to cater for the typical needs of process instrumentation. Multiplexers 'scan' all the data of each type during one 'scan' interval and place the values into memory at specified addresses in a data base which can be accessed by the data-transfer processor in performing its functions of message packaging and data-transfer control. Thus each measurement or contact-state sample transferred to other units in the system, including the operator interface, is at least one and possibly nearly two sample intervals 'old' when received. The sampling interval must be chosen to ensure that the constraints of the sampling theorem are met, and so this delay is not likely to be significant in practical terms.

Another unit, at whatever level, can now request the latest value of any datum to be transferred over the data highway system for use in whatever computation the unit is carrying out. Each such sub-system will take a certain length of time to 'scan' all the inputs of each type: clearly analog signal scan rates will be much slower than discrete signal scan rates since the rate is largely dictated by A/D conversion speed in the former case, and by the multiplexer speed in the latter. The number of signals allocated to each multiplexer or A/D converter therefore determines how frequently each is sampled. This is an important point to remember when comparing one system's performance with another, since the price per signal point is obviously directly proportional to the frequency of sampling. It is always possible to increase the sampling rate of one or two signals by connecting them to two, four or more inputs of the system, so that they are scanned twice, four times or more during one scan period. In this case the sampled values of these particular signals, in the data base, will never be 'older' than a half or a quarter of the scan period. Typically the scan period will be less than a second.

The data-gathering units in a distributed control system usually incorporate output data as well as input, so that the data base will include sampled data values of the analog output control signals to final control elements. Separate multiplexers and D/A converters are required for these output signals.

## 19.6 CLOSED-LOOP CONTROL UNITS

This type of unit sits on the slower 'real-time' data highway, and requests measurement data as necessary from any data-gathering unit. It uses this data, in executing compiled and assembled programs, to calculate the current 'sampled' value of the analog control signal which the appropriate data-gathering unit is outputting to a final control element. The outputting equipment 'holds' the value of the signal between 'samples', so that it, like the measured value signal must conform with the sampling theorem.

It is usual for a controller sub-system to perform the control calculations for a number of SISO controllers on a time-share basis, thus keeping the price down. The same program is executed for each controller in turn, but the variables representing control action terms and the measurement data are changed for each execution. The variables will probably be stored in memory within the controller unit, since they will only be changed occasionally from the operator interface unit. The data, on the other hand, will be requested from the acquisition unit/s as required. The resulting output signal values will be placed in a data base to be accessed on request by the data-acquisition/outputting units. As in the case of the measurement data, a delay will be occasioned by sequencing of the calculations. Typically each controller output will be recalculated 2 or 3 times per second, which for most process control needs is more than adequate.

In the normal way SISO controllers which are linked together into a cascade or ratio system will be within the same sub-system and the output of one controller can be taken from the data base by the processor as the set point of another. There is no reason, however, why the output of any controller anywhere on the plant cannot be used as the set point of the slave controller in a cascade system. The only difference is that regular data transfers between the two sub-systems containing the two controllers will be necessary. Other control functions such as 'ramping' or logical or time-dependent switching for programmed 'batch' control systems can be performed by the same or a second processor within the controller sub-system. One of the greatest advantages of digital control (which includes DDC systems) over analog systems is in accuracy of calculation which can be applied to multiloop and batch control systems.

## 19.7 THE OPERATOR INTERFACE SUB-SYSTEM

The operator interface sub-system provides the means for the process operator to monitor the behaviour of the plant in the traditional way. He does this by 'calling up' displays which consist in general of a fixed background graphic in a form similar to the engineering drawings (such as flow charts and instrument diagrams) which he is used to studying. Superimposed on these graphics are 'live' displays of measured value data and also sometimes of control actions, etc. These displays are generated by compiled and assembled programs held in the system memory. The equipment system manufacturer usually provides a hierarchy of such views, so that the operator can move from an overview of the entire plant to more detailed views of parts of it. In changing the views and in entering new control set points or alarm values, etc., the operator uses other interactive programs which operate with 'menus', etc., in a way which has become standard in personal computers. Other interactive programs provided by the equipment manufacturer enable him to construct or alter the graphic displays himself, although this work is often carried out by the manufacturer as part of the purchase contract. The same graphic displays can be used for historical data reruns.

Other displays are usually provided, which show the status of all measurements relative to alarm and shut-down settings, or give shift reports on alarm/shut-down occurrences.

Operator access to the system is usually from the keyboard which relates to a VDU computer-type screen. Sometimes a touch-sensitive screen, a 'mouse' or a 'light pen' allow the operator to initiate actions such as starting or stopping pumps, without resort to the keyboard. The keyboard itself may comprise a number of 'special' keys which define the operator's actions in process terms, requiring no computer familiarity at all. On the other hand, the keyboard may be a standard computer 'qwerty' device, allowing more complex interaction between man and system. Most systems have both types of keyboard or a combination, in order to cater for the process operator whose functions are restricted to operating the plant facilities, and the engineer who must be able to change control actions, reconstruct displays or even change the structure of the control schemes.

The keyboards of most systems incorporate a number of so-called 'soft keys', which name stems from the fact that the function of any key is determined by 'software' and changes according to the activity being carried out. The bottom or top of the VDU screen is devoted to a related display of the current functions of the soft keys. In this way the operator has a very large number of 'special' keys available to him, many more than could be provided physically.

The VDU display screen is used in two distinct ways. In the first a 'library' of shapes, each of such a size as to occupy a very small part of the screen, enables the engineer to construct background displays in the format of a flow or instrument diagram, with which the operator will be familiar. The screen is typically divided up into 80 columns by 48 rows, giving nearly 4000 locations in each of which a library shape can be displayed. The display is stored in the form of a specification of the shape and its location on the screen, and can be very quickly constructed by a processor when 'called up' by the operator. Onto such displays can be superimposed 'live' data such as measured values. The engineer can construct the required displays or alter them very quickly because of the use of library shapes. He is usually able to construct his own peculiar shapes to add to the library. Thus, even though higher definition displays are possible, they are not usually desirable because of the greater difficulty in construction and also because of the time which the processor would take to construct them on the screen. In operating the plant the operator often changes from one display to another with great frequency, particularly when changing from an 'overview' to a high definition display (see Fig. 19.3).

Each shape is constructed by addressing a set of dots on the screen, which together fill the space allocated. In this way the processor, in building the display, calculates the relative position of each dot within

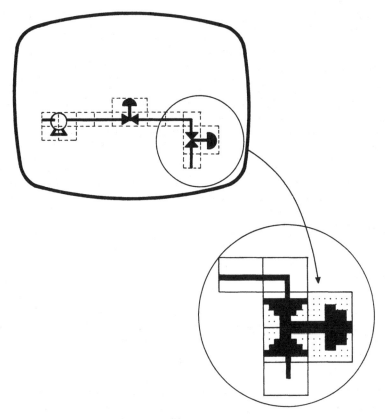

**Fig. 19.3.**

the small area and adds this to a global map reference for the area itself. This is much quicker than having to construct a display from positional data relating to the entire screen.

The second way the display screen is used is in 'high resolution mode' for graphical displays of recorded data. Since such displays leave most of the screen as a common background, it is practicable to define display data on a global basis, which avoids the 'chunky' appearance of the process display, and gives accurate positioning of the relatively few 'dots' to be defined.

The actual scope and hierarchy of displays made available by any particular equipment manufacturer is a matter of his interpretation of the user's needs. It is therefore perhaps a little spurious to describe this aspect of system design here. Any manufacturer will provide very detailed and colourful literature on this aspect of his equipment, whatever else he does or does not provide!

## 19.8 HISTORICAL DATA

One of the most significant advantages of the distributed control system is the facility to store very large amounts of measurement and event data and to perform reruns of past operations long after the events took place. The historical data sub-system will normally sit on the faster data highway. One way this facility can be used is to record all samples of selected measurements over say an hour, and then transfer the 'captured' data to a 'floppy disc' bulk storage device if the operator decides that events took place during that hour that justify post operative analysis. Other ways that the facility can be exploited include regular or 'on-demand' collection of short duration (e.g. 1 min) 'snap-shots', or hourly, shift, daily or monthly averages.

Event history can also be 'captured' by a different process. The time at which an alarm or shut-down signal occurred and was acknowledged by the operator can be recorded by reference to the system real-time clock. Such data-logging can be extended to include operator actions such as changing process set points or changing the configuration of a multiloop control system, for instance.

The historical data can be rerun on the normal operator interface VDU facility in much the same way that TV video recordings can be shown instead of 'live' programs. Since this activity interferes with the normal use of the VDU operator station, it is common practice to provide at least two such stations in order to avoid such interference. One operator station can then be utilised for running historical data or for such 'off-line' activities as constructing displays.

One very useful use of the historical data facility is to provide displays of the 'trend' of (keyboard) selected measurements over a period up to the present. This acts as a direct replacement for paper chart recordings used in earlier systems, and provides great flexibility in recording data from any measurement, over any period of time (subject to the resolution required). Multiple displays provide the facilities once given by multipoint recorders, and the data can be 'captured' for posterity on floppy disc or Winchester mass memory devices.

## 19.9 COMPUTER INTERFACING

The lowest level in the system architecture is concerned with measurement of system parameters, the second level with real-time data processing. The third level is the operating level, and this function can be carried out either by the human process operative or by a 'supervisory' computer. The addition of such a computer to earlier analog instrumentation systems entailed a great deal of complex and hardware special interfacing. Whilst this was not true of centralised DDC systems, it was true that the additional processing load on the single central processor tended to interfere with the basic instrumentation system functions. This is because the demands on processor time of many relatively short real-time tasks related to measurement and control action were basically incompatible with the relatively long supervisory tasks. The use of two computers in a foreground/background configuration was a complex, expensive and inflexible solution to this problem. The distributed processing system manufacturer approaches this problem by providing a standard interfacing system to sit on the high speed data highway. As a result, more than one supervisory computer can be added to the system at any time. Because the optimising or process management processes are carried out by such supervisory computers, independently of the real-time processing functions of the system, they cannot interfere with the latter.

The essential elements of a typical computer gateway unit were shown in Fig. 18.2. This unit provides a data base in its memory which consists of the latest values of all the measurements and other data which the supervisory computer will require in order to carry out such operations as regression analysis for plotting operating graphs, etc. The process of transferring data from any device in the system into the computer interface unit is thus 'buffered' from the process of transferring data from the interface unit into the computer. The two transfer operations can run under totally different protocols and at totally different speeds. The speed at which measurement data is updated in the data base of the interface unit will almost certainly be so much faster than the requirements of the supervisory processing, that sampling delays will be unimportant. The computer/interface communication link is likely to be to a balanced voltage standard such as RS 422, allowing the supervisory computer to be located at some distance from the central control room if required.

## 19.10 ALARM MONITORING

One of the few weaknesses of the distributed control system in comparison to earlier analog systems is its ability to monitor the multiplicity of alarm states common in modern process plant. This stems from the fact that discrete signals (i.e. alarms) are scanned in sequence over an admittedly very short period of time, and the data collected is put into a data base to be accessed by a data-transfer processor when constructing message packages. When an alarm or shut-down signal (particularly the latter) changes state there is often a chain reaction of other alarms/shut-down signals, and it is vitally important to safe operation to establish which signal changed state first. If the signals are scanned by the same multiplexer and changes of state take place on a time scale comparable or even shorter than the scan time, then it is not possible to identify the 'first up'. This is certainly likely to present a problem in the case of vibration alarms on a large compressor, for instance, but may not in other, slower, plant. It is a problem that the user must pay considerable attention to in specifying the system for purchase, and it is the reason why, in many cases such as offshore petroleum facilities, which rely heavily on complex safety shut-down systems, it is now common practice to have an entirely separate safety system.

On the other hand, the distributed control system operator interface is capable of providing a multiplicity of displays which assist the operator by showing the 'historical' relativity of alarm occurrences, displaying alarm states in the context of the graphical displays and with relative importance clearly shown by colour coding. Logical interlocking between alarm states and discrete control actions (starting or stopping pumps, opening valves, etc.) is very easy to implement. Such features as the 'acceptance' of an alarm by the operator by merely touching the VDU screen make a real contribution to the ease of operation and minimising the staff required to operate the plant.

## 19.11 REDUNDANCY, FAULT-MONITORING AND MAINTAINABILITY

As was mentioned at the beginning of this chapter, there must be at least two cables, each run along a different route, for each data highway. This however is far from the end of the requirement for 'redundant' operation. Each sub-system could both transmit and receive signals to/ from both cables simultaneously and compare the messages. This would give a high degree of error protection but would be too expensive in terms of equipment and operating time. There would also be the problem of which signal was correct and which in error: 'voting' systems require at least three channels to give a two out of three decision. Such systems are more appropriate to emergency shut-down systems which will be discussed in the next chapter.

A more appropriate design philosophy for a distributed control system utilises only one cable at a time, switching frequently between the two. Should one cable not be functional when the system switches over, this will be detected and the system will automatically revert to the other cable and output an alarm to the operator saying that there is now no back-up cable. Such switching can be achieved without disturbing normal operation significantly.

This concept of unit redundancy can be applied to the sub-systems themselves, but this is rarely done in a distributed process control system simply because it would be prohibitively expensive. In any case the equipment is inherently very reliable in most respects, and because of its distributed structure the failure of a single sub-system is usually acceptable, provided that replacement units are available on a plug-in basis. Standard units can be kept as spares and can be very quickly 'customised' when plugged in by downloading set point, control action and other 'variable' data from the operator interface system.

Software failures can occur, and have rather different effects from hardware failure as a rule. It may not be immediately obvious, for instance, that a processor has become locked into a 'loop' and is not performing any of its tasks. The concept of the 'watchdog' overcomes this problem in the case of the 'real-time' units. An independent 'countdown timer', which the processor must reset each time it finishes 'servicing' a task program (such as calculating an updated control output for instance), will raise an alarm if it 'times out'. Thus if the processor fails, for any reason whatever, to complete a task and return to the function of resetting this timer, an alarm is automatically raised. The watchdog is powered by a battery back up of its own and is truly a fail safe device; its functioning does not depend in any way on other equipment in the sub-system.

Other system fault conditions can also be monitored as part of the normal system operating functions, and this provides a considerable benefit which was not possible with analog systems in the past. Similar diagnostic functions were of course possible with DDC systems, but these were much more vulnerable to total breakdown due to the centralisation of functions which is unavoidable with a single central processor.

How easy a system is to maintain, how quickly faults are detected and corrected, the cost and variety of the spares held and the level of expertise required of maintenance staff are all questions of the greatest importance to the user of a distributed process control system. The fault-monitoring system is the key to the answer to all these questions. Faults may be intermittent or continuous in nature and a fault may not prevent the sub-system in which it occurs from carrying out its normal functions. Such intermittent or non-critical faults must be detected, since they usually lead to critical faults eventually. Detection can be achieved by building suitable 'checks' into the operating software of the sub-system and by watchdogs. Other faults may cause the sub-system to generate errors in the results of its processing. Such errors must be detected by different 'checks' built into the data-processing software. Catastrophic failure of a sub-system is very easy to detect of course. The fault-monitoring system should be capable of analysing the errors and/or faults detected and determining the necessary corrective action. The variety of sub-systems must be kept to a minimum which may involve some measure of redundancy. In this respect the facility to change the functions of a sub-system by reprogramming its memory or by replacing a PROM-based memory is extremely useful.

Fault tolerance is of the greatest importance in any system. If a unit or sub-system fails, it is sometimes possible for other similar units sitting on the same data highway to perform its functions instead, provided the fault-monitoring system is able to instruct them automatically to do so. This may entail some slowing up of the system, or alternatively it may be possible for the back-up units to be relieved of some of their less essential functions temporarily. Such overall organisational management functions are usually vested in an operator interface sub-system, and may require the operator to make decisions in some cases. Obviously such back-up action is only temporary in nature, and the faulty unit must be replaced as soon as possible. Lastly, correction of any fault should entail no more than replacement of the faulty sub-system; repair is usually effected by the manufacturer who alone has the necessary testing facilities.

## 19.12  SCADA

System Control And Data Acquisition (SCADA) is a name which is usually reserved for extensive systems related to on- or offshore petroleum or gas-gathering plant. Distances between bore wells and centralised facilities are measured in many kilometres, and often the facilities are connected by pipelines over even greater distances. Communication links in such systems will in many cases be UHF radio or line-of-sight radio (radar). At each 'node' of the system there will be some measurement and regulatory control systems and a local operational centre. There will probably be only a single data-gathering and output sub-system. Star rather than data highway architecture is obviously most appropriate for communications between outlying nodes and the central facility, where there will probably be a distributed system. Because the star communication links are long, FSK signalling will almost certainly be used: TDM or FDM multiplexing techniques will be used over these long-distance 'dedicated' links.

The reader will appreciate that microprocessors in the form of VLSI chips will certainly be used to construct today's equipment systems for this type of application. However the communication problems and the architecture of such systems is considerably different from the type of distributed control system appropriate to a refinery or paper mill, for instance. The major problems will be concerned with signal degradation and communication link performance, and the speed of data transfer will be of an order lower than is common in the other type of system. Different manufacturers tend to specialise in one or other types of equipment system, and the user is likely to find himself dealing with the problems of interfacing the different manufacturers' equipment systems. This is perhaps the most difficult part of a user or design contractor's work and requires a knowledge of the subject matter introduced in the earlier chapters of this book.

## 19.13  CONCLUSION

The reader will appreciate that each equipment manufacturer has his own solutions to the many design problems, and the many compromises between cost, complexity and safety involved in the design of any distributed control system. It is for the user to understand these factors in sufficient detail to be able to judge which system suits his needs. The intention of this book is to present the reader with sufficient knowledge to enable him to ask the appropriate questions and thus to make valid judgements based on the answers. Most manufacturers provide extremely comprehensive and well-illustrated technical literature which the reader should' be able to follow as a result of reading this book. It would therefore be inappropriate to try, in this book, to review any particular system, or all systems, in the sort of detail best left to the equipment manufacturer to describe.

# CHAPTER 20

# *The Microprocessor at the Closed-Loop Level*

## 20.1 'INTELLIGENT' TRANSDUCERS

The standard 4–20 mA analog signal which is generated by most analog transducers has to be converted into digital form by the data-acquisition sub-system (Section 15.7). This adds very greatly to the complication of data-gathering sub-systems. Microprocessors are now being used in the transmitter itself to enable samples of the measured value to be generated directly in digital code by the force-balance mechanism, thus removing the need for A/D conversion at the data-acquisition stage, and consequently speeding up the scanning process. Compromises have to be made between the precision with which the analog signal can be converted to digital code on the one hand and speed of scanning on the other, in the design of data-acquisition sub-systems. By generating digital sampled value signals at the transducer these compromises can be avoided. Exactly the same advantages could be obtained, though with less flexibility, using individual A/D converters in the data-acquisition system for each signal. There are, however, several other advantages to be gained from the introduction of processing power to the transducer:

1.  It is relatively simple to apply corrections to the signal, such as square root extraction (for flow rate signals generated by pressure difference primary devices) and pressure and temperature corrections etc. at the transducer itself. Any constants used in the software programs which are stored in the transducers memory, and which are used in making these corrections to each sample value, can be downloaded from a central operator interface sub-

system. Thus the scaling of the transducer can be checked and altered remotely.

2. Automatic calibration can be effected by downloading a calibration signal, either from a central operator interface or a hand held calibrator. From this signal the processor can calculate new calibration factors to be used in the execution of its software programs at each sample interval.

3. Fault monitoring can be extended to cover the transducers in a distributed control system.

4. The transducer itself can sense when the measured value exceeds a preset alarm level and this status can be transmitted along with each sampled value. Taken together with the elimination of A/D conversion and consequent increase in scanning speed this may well offer a partial answer to the 'first up' detection problem.

A possible future development may well be the elimination of data-acquisition sub-systems altogether. Each transducer would then 'sit' directly on the data highway, and would have two processors, the second dedicated to communication handling. A separate communication handler is required because the transducer will have to be capable of working with a number of manufacturers' distributed systems, and each will have its own protocols and interfaces. By changing the communication handler alone the transducer manufacturer will be able to provide a product to suit each without undue difficulty. Even if this were not necessary it is unlikely that a single processor would realistically be expected to time share the instrument signal processing tasks outlined above with the communication tasks. It is probably true that, at the present time, the cost of hardware to implement this development is still too high. The problems associated with signal degradation and interference are quite different for analog and digital signal transmission, and since communication between 'field' transducers and the distributed system itself are inevitably subject to the worst conditions in this respect, this is a vital consideration in the development of digital, and therefore intelligent, transducers. The balance of advantages is not necessarily in favour of digital signal transmission in this respect! Use of optical-fibre light transmission would of course shift this balance substantially in favour of the digital transducer. However the transducer must still be powered, and explosion and fire hazard considerations preclude the use of batteries for this purpose: hence a power-supply cable would be needed in addition to the fibre-optic cable, making this an expensive option. Solution of this problem would make the intelligent digital transducer very attractive indeed!

## 20.2 'INTELLIGENT' SISO CONTROLLERS

One of the greatest advantages of digital control is the accuracy of calculation that can be brought to the various functions associated with multiloop control systems. These functions include, but are not limited to, adding, subtracting, multiplying, dividing, ratioing and scaling of signals. It is also very easy to provide lead-lag compensation, dead-time compensation, and such non-standard control algorithms as inverse derivative, squared-error proportional action, etc. Discrete control actions such as starting and stopping integral control action in order to prevent integral saturation or 'wind-up' in batch control, switching between master controllers in a cascade system, bumpless transfer, etc., are also very easily provided as standard software programs to be stored in the memory of a control sub-system and called up as part of the customisation of such standard systems to suit the particular requirements of a specific control system. Such customisation is usually effected at the operator interface using other interactive programs supplied by the equipment manufacturer which talk the user through the customisation procedure in the same way that he is talked through the procedure of setting control actions and set points. The customisation is then downloaded to the controller sub-system. Often a separate operator interface is provided for this type of system customisation so that the plant operators cannot restructure the control system; alternatively some means of key or password access protection is provided.

## 20.3 INTELLIGENT FINAL CONTROL ELEMENTS

It is common practice to fit a 'positioner' to a final control element or control valve in order to isolate its dynamic behaviour from the process controller. In effect the positioner is a slave controller and the process controller a master. The reason for fitting a positioner has much to do with an unsatisfactory installed valve characteristic which cannot always be avoided because of the conflicting requirements of rangeability and linearity. However, one thing the positioner does not do is compensate for the incorrectly installed characteristic, with the result that the loop gain of the process controller has always to be a compromise to suit all operating conditions. An 'intelligent' positioner, using a microprocessor, can readily characterise the output to the final control element in such a way that the installed characteristic of the valve/positioner as 'seen' by the process controller is linear, even though the characteristic of the valve itself does not give installed linearity. In this way the gain of the process controller can be optimal for all operating conditions, with overall benefit to control quality. Such techniques are currently being applied to hydraulic control mechanisms but the process industry does not yet seem to have realised the advantage.

## 20.4 'REAL-TIME' OPERATION

All the foregoing control functions are carried out discreetly. A new value of each analog control output, computed at frequent intervals from the latest sampled data, replaces the previous value. The scheduling of these data-gathering, discrete computations and A/D conversions is regulated by a 'real-time' clock. An 'executive' program, run continuously by the sub-system, schedules when each task is run, as shown in Fig. 20.1(a). The executive program may operate a system of priorities so that the sub-system can carry out tasks that take appreciable time but do not have to be done very often, and tasks which take only a very short time but must be carried out frequently at

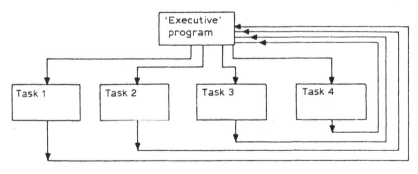

Fig. 20.1(a)

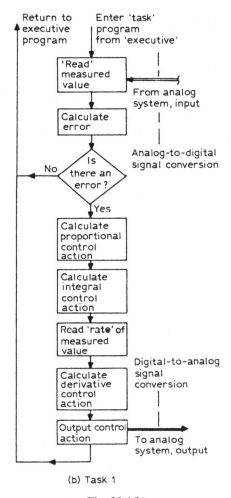

(b) Task 1

**Fig. 20.1(b)**

precisely determined times (e.g. updating measured-value or output-signal data bases).

One of the 'tasks' might be the computation of a new control output as illustrated by the 'flow chart' in Fig. 20.1(b) which shows the sequence of operations the processor will have to carry out to execute this task. In fact a sub-unit which performs the functions of say six SISO controllers will have to carry out six identical tasks in strict sequence, each using the same program, but different data (measured values) and different constants (set points and control action gains). Provided the intervals between the execution of one control output computation and the next for each SISO controller is not too small (say one second), then there will be considerable unused computing time between the completion of one such task and the time to start the next.

Tasks which do not have to be completed quickly, or which are not real-time dependent can be partly executed during the available unused computing time. This is achieved by setting 'flags' in the program which carries out such extended tasks and by storing all the data at the point reached by the program when the flag is set. The processor can then return to the task and resume the computation at the point it left it, when the next unused computation interval occurs. This mode of operation is known as 'multitasking'.

## 20.5 SEQUENCE CONTROL

The digital processor is designed to carry out logical operations, and it is therefore the ideal device for sequence-type control such as is encountered in batch operations. Each step in the sequencing of a batch operation is initiated by a discrete input from a process sensor such as a temperature 'switch' or on a real-time basis. Sometimes a part of the sequence of operations is repeated when a given state or time is reached, and this repetition may itself be conditional on the attainment of some other condition. All this can readily be incorporated into a program and the system manufacturer usually provides a suitably specialised high level language compiler for the purpose. Traditionally the sequence program is defined in the form of a 'ladder' diagram as shown in Fig. 20.2 because, for the most part, the output signals form part of electrical engineering circuits. The equipment manufacturer may therefore provide a 'constructor' program which enables the process engineer to construct his process sequence in terms that are familiar to him (not necessarily a ladder diagram). The 'source code' which is produced in this way is then compiled into 'executable code' which the processor can run. This may seem unnecessarily complex, since the process sequence control has all the same elements as a computer program; however, this method of construction enables the process engineer to make changes to the sequence control at any time by changing the source code and having a new executable code compiled and downloaded. If the sequence control program is written in a conventional high-level

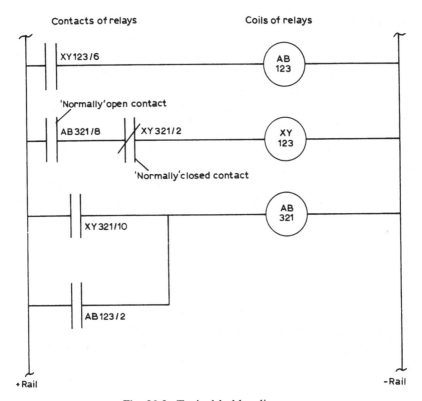

Contacts of relays                    Coils of relays

**Fig. 20.2.** Typical ladder diagram.

language this will not be so easily achieved since the program will have to be redesigned in some way to reflect the new sequence control requirements.

Sequence control programs are often run by a sub-system which, at the same time, carries out a number of SISO continuous control algorithms. The executive program 'looks' to see if any sequence input has changed state after completion of each SISO controller computation. If one has, or it is time to 'step on' in the sequence, the necessary outputs to start/stop pumps, open/shut valves, etc., will be implemented within the task program. In fact it is usually possible to incorporate several such sequence tasks within the executive program of a typical sub-system.

## 20.6 GAIN SCHEDULING

Distributed processing architecture, however, makes it possible to provide complex higher-level facilities, which would have overloaded any single central processor, by adding an additional sub-system to handle the extra computation independently, and download to the SISO level control sub-system changes of value to such controller constants as set points and control action factors. In most cases the equipment manufacturer will provide a sub-system which is essentially a microcomputer with the necessary interfacing to the common data highway for these purposes. He may provide a special high-level language and software environment which amongst other things will make it easy for the user to call for the changes in constants that such higher-level techniques require.

An example of such a control technique is 'gain scheduling' which was developed for flight control systems in the aviation industry and has much to recommend it to the process industry. Proportional gain in a closed loop is varied (scheduled) according to the measured value of an appropriate operating condition at any particular time. Derivative action can be considered as varying the proportional action gain in proportion to a measurement (made by the controller itself) of rate of change of the measured variable. Rate of change of a variable is of course an independent variable: thus derivative control action can be considered as a form of gain scheduling. An operating parameter which correlates with optimal gain of a control loop must be identified. Gain is then 'scheduled' according to the measured value of this parameter.

A good example of gain scheduling is to be found in the 'gauge' control of metal sheet rolling processes. The 'gauge' or thickness of the sheet, as it emerges from the rollers, is measured. This measurement is used in a closed-loop feedback control mechanism, the final control element which contains the hydraulic 'jack' which applies pressure to the rollers (see Fig. 20.3). The proportional action factor of this control system is the ratio of the change of 'gauge' to the error signal (the difference between the required gauge and the measured gauge). This in turn is given by the gain of the jack mechanism or the change in force applied per unit change in control signal, multiplied by the 'strength' of the sheet. The strength of the sheet depends, in turn, on its width and its modulus of compressive stress, both of which are parameters which can be measured on material before it enters the mill. These parameters can

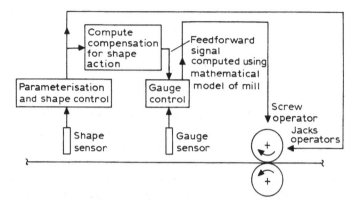

**Fig. 20.3.** Use a feedforward control to minimise shape/gauge interaction.

be fed into the control system to 'schedule' the proportional action gain of the controller so that it remains constant despite any changes in material width and strength.

## 20.7 SELF-TUNING CONTROLLERS

As can be seen from the block diagrams in Fig. 20.4 the self-tuning controller differs from the gain-scheduled controller only in as much as the former's gain is modified by what might be described as an 'outer' loop, the measured value of which is some measure of the response of the regulator (SISO) loop itself. This outer loop acts on the gain of the SISO controller rather than its set point, but is otherwise related to it in the same way as a master controller in a cascade system. Thus the gain is varied in closed loop, unlike gain scheduling, which is a form of feedforward control.

For the outer control loop of a self-tuning controller to be able to measure in some sense the response of the regulator loop itself, the loop must operate in unsteady state, which is to say that the measured variable must be oscillating to some extent. This is not necessarily a bad thing as it means that control is slightly underdamped and therefore will respond quickly to upsets. There are many ways in which the response can be assessed, all of which involve estimating the transfer function of the closed-loop system. A discussion of the mathematical techniques involved is beyond the scope of this book but it is perfectly feasible to encapsulate these techniques in a compiled program which can be run

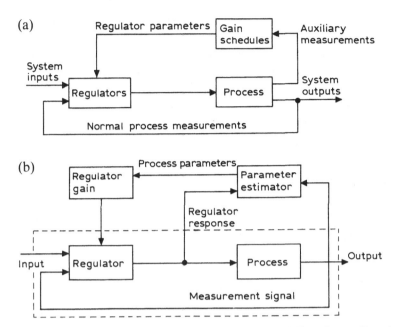

**Fig. 20.4.** Block diagrams of (a) a system with gain scheduling; (b) a self-tuning regulator (STR).

by the processor of a suitable sub-system as described above. Much research is being carried out into suitable mathematical techniques for different applications. In process control, however, measured values change only slowly in most cases, and there are self-tuning process controllers available commercially. Provision of a standard algorithm for use within a distributed control system would appear to be a relatively simple matter!

## 20.8 SOFTWARE 'DOWNLOADING'

Perhaps the most startling demonstration of the flexibility which the microprocessor has brought to the industrial instrumentation system is the ability to almost totally 'revamp' the functions of the system within the constraints imposed by the system architecture only. Earlier hard-wired systems were almost totally inflexible in this respect and the early DDC computer systems, though much more flexible, fell a long way short of the 'distributed' systems of today.

Each microprocessor in a system has its own operating system which is stored in ROM and which contains the software (or 'firmware') associated with the functions it is intended it will carry out. These functions have been described in some detail earlier in this chapter. It is easy to see that, if the operating program for any particular micro-processor is overwritten, its functions are completely changed. Thus, not only can the control action settings (set points, proportional band, etc.) be changed from the operator interface, but the control actions themselves and even the interconnections between individual SISO controllers (ratio, cascade, etc.,) can be changed, provided only that the necessary measurements are available to the system. Within limits it is even feasible to make such changes whilst the process continues in operation, since the output of any controller is its integral action term in steady state and this can readily be 'frozen' temporarily.

The ROM memory in which the operating system is held must be reprogrammable under specified conditions, though it must be certain, nevertheless, that it cannot be 'corrupted' during normal operation. The emergence of special types of ROM such as the EPROM, or electrically programmable read only memory, has made it possible to engineer into any distributed processing system a mode of operation known as the 'downloading mode'. In this mode all the operating programs are transferred over the communication links of the system, as data. Care must be taken that this data is in no way corrupted, but this is easily done by retransmitting all data transfer messages back to origin. If what was sent and what is retransmitted back do not tally exactly, normal operation of the system will not be resumed until the error has been found and corrected.

The value of such flexibility is immense: it means that a large part of the design of control systems can be considered as 'software' rather than hardware as used to be the case not so long ago. It is probably of even more significance in the case of the advanced control techniques introduced in the next chapter and in emergency shut-down systems (Chapter 22), though it also introduces new problems in terms of the security of fail-safe operation.

# CHAPTER 21

# *Advanced Control Techniques*

## 21.1 PLANT SYSTEM 'STATES'

The system states, temperature, overhead flow, etc., represent coordinates of the system space, in this instance the plant operation. The coordinates, however, are not independent (orthogonal in geometrical terms). We can illustrate what this means in the case of a two-dimensional system as shown in Fig. 21.1.

In the left-hand figure the system operating point is defined in terms of the orthogonal coordinates or states $x$ and $y$. There is a *direction* in 'space' in which the system operating point can change in one case only affecting $x$ and in the other only affecting $y$ ($x$ and $y$ might represent temperature and pressure, say). In the right-hand figure the operating point is defined in terms of non-orthogonal coordinates and no matter in what direction the system operating point moves, both $x$ and $y$ states are affected. Interpreting this in terms of process system states, we must realize that it is no more than a fortunate coincidence if the 'natural' measured states of a process happen to be independent. Indeed, it is very rarely that this is the case; however, often the degree of dependence is

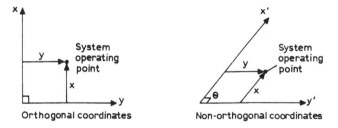

**Fig. 21.1.**

355

small ($\theta$ in the right-hand diagram approaches 90° orthogonality) and can be ignored for practical purposes. The selection of pairs of measured manipulated variables so as to minimise interactions can be made from knowledge and experience of the physical relationships of the process. This may not always produce the optimum system and, for a large system having a great many dimensions, may not even be possible. Bristol has shown how the selection can be made in a logical fashion either from the available process design data or from the results of systematic testing on site. An array or table is constructed, having as its columns the inputs to the system (whether manipulable or not) and as its rows the measured variables of the system. Into each space in this table is inserted the 'open-loop gain' between the relevant measured value and the corresponding input. The open-loop gain is defined as the measure of the affect of a change in the input on the measured value, assuming all other variables do not change.

Thus:

| | $u_1$ | $u_2 \; \ldots \; u_n$ |
|---|---|---|
| $x_1$ | $a_{11}$ | $a_{12} \; \ldots \; a_{1n}$ |
| $x_2$ | $a_{21}$ | $a_{22} \; \ldots \; a_{2n}$ |
| . | . | . . |
| . | . | . . |
| . | . | . . |
| $x_n$ | $a_{n1}$ | $a_{n2} \qquad a_{nn}$ |

is an 'array' which defines the relative gains between the $n$ inputs $u_1$, $u_2, \ldots, u_n$ and the $n$ measured values $x_1, x_2, \ldots, x_n$, $a_{ij}$ being the gain between $u_j$ and $x_i$. For convenience, this array can be rearranged in the following manner without changing the meaning in any way. It will be appreciated that this equation represents a set of equations of the form

$$x_n = a_{n1}u_1 + a_{n2}u_2 + \ldots + a_{nn}u_n$$

$$\begin{bmatrix} x_1 \\ x_2 \\ x_n \end{bmatrix} = \begin{bmatrix} a_{11} & a_{12} & \ldots & a_{1n} \\ a_{21} & a_{22} & \ldots & a_{2n} \\ a_{n1} & a_{n2} & \ldots & a_{nn} \end{bmatrix} \begin{bmatrix} u_1 \\ u_2 \\ u_n \end{bmatrix}$$

*Note:* The columns of $x$ and $u$ terms are referred to as 'vectors' and denoted **x** and **u**.

The open-loop gain of any state $x_i$ to any input $u_j$ represents the *steady-state* change in $x_i$ which will result from a unit change in the input variable $u_j$ in normal operation. This figure will normally be obtainable from a *steady-state* design such as a chemical engineer must produce. If we can define some overall measure of the effect of the input vector **u** on the state vector **x** and then eliminate any particular pair of state/input variables $x_i$ and $u_j$ from these vectors and obtain the corresponding overall measure without them, we can assess the degree of coupling of this pair of variables. If the measure is reduced by exactly the inverse of the steady-state 'open-loop gain' then we know that the two variables under consideration are independent. We can illustrate this by considering the three-dimensional system shown in Fig. 21.2. The states $x_1, x_2$, and $x_3$ have been increased by changes in inputs $u_1, u_2, u_3$, and the overall measure of effect of the change in the vector **u** is the volume change represented by $(x_1' \cdot x_2' \cdot x_3') - (x_1 \cdot x_2 \cdot x_3)$. In the special case where the states $x_1, x_2, x_3$ are all orthogonal, we now consider the two-dimensional subsystem $x_1, x_2$, ignoring $x_3$ (and $u_3$) in Fig. 21.3.

The overall effect of the change in vector **u**, is now the area shaded. If

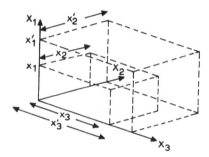

**Fig. 21.2.**

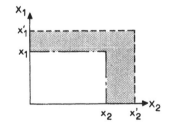

**Fig. 21.3.**

we multiply this area by $(x_3' - x_3)$ we obtain the total effect again. In matrix terms this two-dimensional subspace is obtained by eliminating the row and column containing the element $a_{jk}$:

$$
\begin{bmatrix} a & 0 & 0 \\ 0 & b & 0 \\ 0 & 0 & c \end{bmatrix} \quad \text{becomes} \quad \begin{bmatrix} b & 0 \\ 0 & c \end{bmatrix}
$$

The same principles can be extended to the non-orthogonal case where the states are dependent. In such a case each $u_j$ has a component effect on each variable $x_1, x_2, x_3$ and if we eliminate the effect of, say, $u_2$ on $x_2$, we only eliminate a component of the total effect of $u_2$ on the vector $\mathbf{x}$. Thus the 'volume' $(x_1 x_2 x_3)$ is not reduced by the whole amount represented by the effect of the input $u_2$.

The measure of overall effect (volume) of the input vector on the state vector is given by the determinant of the matrix (which is literally the volume in the case of a three-dimensional system and area in the case of a two-dimensional system). Hence the relative overall effect of the complete input vector to the reduced subspace is given by

$$
\psi_{ij} = \frac{A_{ij}}{|A|}
$$

where $|A|$ is the determinant of $[A]$ and $A_{ij}$ is the determinant of the reduced matrix $w$ with input $(u_j)$ and system state $(x_i)$ removed. The matrix whose elements are these factors is known as the relative gain array RGA. If we obtained a matrix such as this

$$
\begin{bmatrix} x_1 \\ x_2 \\ x_3 \\ x_4 \end{bmatrix} = \begin{bmatrix} \psi_{11} & \psi_{12} & 0 & 0 \\ \psi_{21} & \psi_{22} & 0 & 0 \\ 0 & 0 & I & 0 \\ 0 & 0 & 0 & I \end{bmatrix} \begin{bmatrix} u_1 \\ u_2 \\ u_3 \\ u_4 \end{bmatrix} \quad \text{or} \quad \mathbf{x} = [\text{RGA}] \cdot \mathbf{u}
$$

for the system under consideration we would know that $x_3/u_3$ and $x_4/u_4$ are independent pairs which can be controlled by single-loop controls, thus reducing the size of the interactive system. The matrix this produces can be made more like that which was found to describe the boiler system earlier, by rearranging the pairing of the input and state variables in such a way that the largest terms in the matrix lie on the 'leading' diagonal. In doing this we are minimising the system interactions as far as is possible. It can now be seen that a control system comprising only single loops can only implement the diagonal terms of such a matrix of relationships, and ignores the terms which are not on the leading diagonal and which 'describe' the system interactions.

The matrix generated by Bristol's technique constitutes a steady-state model of the system, in that it describes in steady-state terms the influence that each input exerts over *all* the state variables or dimensions of the system.

## 21.2 COMPENSATING FOR PROCESS INTERACTIONS

The basis for all process system design has always been to identify the variables which need to be regulated, then to choose from those that can be manipulated, one which most strongly affects each of the variables to be regulated. This technique, of course, ignores the fact that there is rarely only one manipulatable variable which can affect the variable to be controlled. In reality there is usually more than one variable which, if manipulated, will strongly influence any particular measured variable. This fact is of course at the bottom of most bad process control design. The facility to restructure the control loops, which both DDC and distributed control systems provide, enables the engineer to rectify bad process control design after plant start up. However, there are some cases in which no one-to-one pairing of control variables gives a good design: the measured variable is almost equally affected by changes in more than one manipulatable variable!

Using the relative gain array matrix as a model of system behaviour,

$$x = [RGA] \cdot u$$

hence

$$u = [RGA]^{-1} \cdot x$$

Thus the inverse of the RGA matrix can be used to compute a 'set' of incremental changes in the set points of slave SISO controllers for each of the manipulated variables in the vector **u**. Implementation of this matrix algorithm on a sampled data basis in a separate microprocessor sub-system constitutes a 'multivariable master controller' which regulates the process states in a decoupled sense.

The following developments demonstrate the design of a steady-state decoupling system for a simple blending control system typical of the digestion 'liquor' system found in the Bayer process for producing alumina. Alumina is the raw material from which aluminium is produced by smelting and is itself extracted from bauxite ore by digestion in a caustic 'liquor' and subsequent crystallisation. After crystallisation the liquor is returned to the beginning of the process as 'old liquid' (OL). During the process, water used to wash filters, etc., dilutes some streams of liquor, which are collected as 'dilute liquor' (DL) and blended with the OL to become 'strong wash' (SW) liquor. New liquor is produced in 'kiers': this is more concentrated (in $Na_2O$) and is called 'kier liquor' (KL). The blend of OL/DL forming SW can be regulated and the blend of SW with KL can also be regulated, independently, giving two 'manipulatable' variables, the *desired values* of which can be manipulated so as to regulate the concentration of the liquor in terms of 'free' (free radicals) soda ($Na_2O$) on the one hand and 'free' alumina ($Al_2O_3$) on the other. The optimal operation of the crystallisation process depends on regulating *both* these variables and so a control system is required which decouples the action of two conventional 3-term controllers to achieve independent steady-state regulation of both free soda and free alumina.

The diagram in Fig. 21.4 shows that the dilute liquor (DL) stream and the kier liquor (KL) stream are both 'wild', whilst the old liquor (OL) and strong wash (SW) (which is the sum of the OL and DL streams) are regulated. Since any control action applied to the SW stream through $CV_1$ must inevitably have an effect on the controller $FC_2$ in the OL stream, there is an obvious need for decoupling in this system. The interactions which need to be decoupled are a direct result of the process requirements and cannot easily be avoided in this instance.

The following shows how the relative gain array is determined from the process data:

|  | | Units of flow | $Al_2O_3$ g/litre | $Na_2O$ g/litre |
|---|---|---|---|---|
| New liquor (NL) | | $(1 + a + b)$ | 135 | 140 |
| Kier liquor (KL) | | 1 | 170 | 155 |
| Old liquor (OL) | } SW | $a$ | 75 | 145 |
| Dilute liquor (DL) | | $b$ | 100 | 100 |

from which:

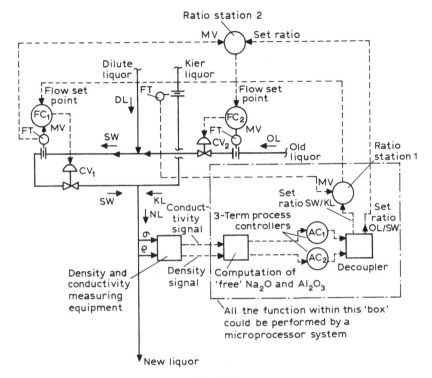

Ratio station 2

MV    Set ratio

Dilute liquor    Kier liquor

Flow set point    Flow set point

FC₁    FC₂

FT    MV    DL    FT    MV    OL

SW    CV₂    Old liquor

CV₁    Ratio station 1

SW    KL    Conductivity signal    Set ratio SW/KL    MV    Set ratio OL/SW

NL    3-Term process controllers    AC₁    Decoupler

σ    AC₂

ϱ

Density and conductivity measuring equipment    Density signal    Computation of 'free' Na₂O and Al₂O₃

All the function within this 'box' could be performed by a microprocessor system

New liquor

**Fig. 21.4.**

for Al$_2$O$_3$    $135(1 + a + b) = 75a + 100b + 170$

for Na$_2$O    $140(1 + a + b) = 145a + 100b + 155$

Hence $a = 0.3400$ and $b = 0.4175$.

*Process Gains*

The first manipulable variable will be the ratio of the flow rates of SW and KL. Hence the process gain: g/litre Al$_2$O$_3$ in NL divided by the percentage change in SW/KL ratio, is calculated as follows:

Al$_2$O$_3$ in SW $= [(75 \times 0.3400) + (100 \times 0.4175)]/0.7575$

$= 88.7863$ g/litre

Al$_2$O$_3$ in KL $= 170$ g/litre

$$\text{Al}_2\text{O}_3 \text{ in NL } = [170 + (88\cdot7863 \times 0\cdot7575)]/1\cdot7575$$
$$= 134\cdot9961 \text{ g/litre}$$

Assuming a $-1\%$ change in the SW/KL ratio:

$$\text{Al}_2\text{O}_3 \text{ in NL becomes } \frac{170 + (88\cdot7863 \times 0\cdot7575 \times 0\cdot99)}{1 + (0\cdot7575 \times 0\cdot99)}$$

$$= 135\cdot1961 \text{ g/litre}$$

Hence a $-1\%$ change in SW/KL ratio produces:

$$\frac{100(135\cdot1961 - 134\cdot9961)}{134\cdot9961} = +0\cdot148\% \text{ change in g/litre Al}_2\text{O}_3 \text{ in NL}$$

Therefore the process gain $= -0\cdot148$.

Similarly, the process gain: g/litre $\text{Na}_2\text{O}$ in NL divided by the percentage change in SW/KL ratio is calculated thus:

$$\text{Na}_2\text{O in SW } = [(145 \times 0\cdot3400) + (100 \times 0\cdot4175)]/0\cdot7575$$
$$= 120\cdot1850 \text{ g/litre}$$

$$\text{Na}_2\text{O in KL } = 155 \text{ g/litre}$$

$$\text{Na}_2\text{O in NL } = [155 + (120\cdot1850 \times 0\cdot7575)]/1\cdot7575$$
$$= 139\cdot9890 \text{ g/litre}$$

Assuming a $-1\%$ change in SW/KL ratio as before, $\text{Na}_2\text{O}$ in NL becomes:

$$\frac{155 + (120\cdot1850 \times 0\cdot7575 \times 0\cdot99)}{1 + (0\cdot7575 \times 0\cdot99)} = 140\cdot074 \text{ g/litre}$$

Hence the process gain $= -0\cdot061$.

The second manipulable variable will be the ratio of old liquor to total strong wash flow rates.

The process gain: g/litre $\text{Al}_2\text{O}_3$ in NL divided by the percentage change in OL/SW ratio is calculated as follows:

The design value of this ratio is $0\cdot340/0\cdot758 = 0\cdot4485 = R$

$$\text{Al}_2\text{O}_3 \text{ in SW } = 88\cdot786 \text{ g/litre}$$

$$\text{Al}_2\text{O}_3 \text{ in KL } = 170\cdot000 \text{ g/litre}$$

$$\text{Al}_2\text{O}_3 \text{ in NL } = 134\cdot996 \text{ g/litre}$$

Assuming a −1% change in the OL/SW ratio:

$$R = (0.99 \times 0.4488) = 0.444\,01$$

then $Al_2O_3$ in SW becomes:

$$\frac{75(0.7575 \times 0.444\,01) + [0.7575 - (0.7575 \times 0.444\,01)] \times 100}{0.7575}$$

$$= 88.90 \text{ g/litre}$$

Hence $Al_2O_3$ in NL = $[170 + (88.90 \times 0.7575)]/1.7575 = 135.045$.

Hence the process gain = $-0.036$.

For $R = 0$, i.e. $-100\%$ change, the process gain = $0.036$. Hence this control action is linear. Similarly, therefore, the process gain: g/litre $Na_2O$ in NL divided by the percentage change in OL/SW ratio is calculated as follows:

$$Na_2O \text{ in SW} = 120.185 \text{ g/litre}$$

$$Na_2O \text{ in KL} = 155 \text{ g/litre}$$

$$Na_2O \text{ in NL} = 139.989 \text{ g/litre}$$

Assuming a $-100\%$ change in OL/SW ratio, $R$ becomes zero. Hence:

$$Na_2O \text{ in SW} = [(145 \times 0) + (0.7575 \times 100)]/0.7575$$
$$= 100 \text{ g/litre}$$

$$Na_2O \text{ in KL} = 155 \text{ g/litre}$$

$$Na_2O \text{ in NL} = 155 + (0.7575 \times 100) - 1.7575$$
$$= 131.2856 \text{ g/litre}$$

Therefore the process gain = $+0.062$.

It is now possible to form the open-look process gain array, thus:

| | Ratio flow SW/KL | Ratio flow OL/SW |
|---|---|---|
| g/litre $Al_2O_3$ | −0.148 | −0.036 |
| g/litre $Na_2O$ | −0.061 | +0.062 |

Taking the array for a two input/two output system as:

|                              |         | Output to SW/KL ratio station $C_1$ | Output to OL/SW ratio station $C_2$ |
|------------------------------|---------|:-----------------------------------:|:-----------------------------------:|
| $Al_2O_3$ control action     | $M_1$   | $w$                                 | $x$                                 |
| $Na_2O$ control action       | $M_2$   | $y$                                 | $z$                                 |

$$\begin{bmatrix} M_1 \\ M_2 \end{bmatrix} = \begin{bmatrix} w & x \\ y & z \end{bmatrix} \begin{bmatrix} C_1 \\ C_2 \end{bmatrix}$$

The effect of any steady-state changes in $C_1$ and $C_2$ on $M_1$ are:

$$\delta M_1 = w\delta C_1 + x\delta C_2 \tag{1}$$

and similarly:

$$\delta M_2 = y\delta C_1 + z\delta C_2$$

*Decoupling Matrix*

A second array is now defined such that it enables a set of $\delta C_n$ to be computed which will influence only one of the measured variables:

|                      | $\delta M_1$ | $\delta M_2$ |
|----------------------|:------------:|:------------:|
| $\delta C_1$         | $p$          | $q$          |
| $\delta C_2$         | $r$          | $s$          |

$$\begin{bmatrix} \delta C_1 \\ \delta C_2 \end{bmatrix} = \begin{bmatrix} p & q \\ r & s \end{bmatrix} \begin{bmatrix} \delta M_1 \\ \delta M_2 \end{bmatrix}$$

Then

$$\delta C_1 = p\delta M_1 + q\delta M_2$$

and                                                                                        (2)

$$\delta C_2 = r\delta M_1 + s\delta M_2$$

Combining 1 and 2:

$$\delta M_1 = w(p\delta M_1 + q\delta M_2) + x(r\delta M_1 + s\delta M_2)$$
$$\delta M_2 = y(p\delta M_1 + q\delta M_2) + z(r\delta M_1 + s\delta M_2)$$

Simplifying

$$1 = wp + wq\frac{\delta M_2}{\delta M_1} + xr + xs\frac{\delta M_2}{\delta M_1} \tag{3}$$

$$1 = yp\frac{\delta M_1}{\delta M_2} = yq + zr\frac{\delta M_1}{\delta M_2} + zs$$

The requirement that only one measured variable must be affected dictates that:

$$\delta M_1 = 0 \quad \text{when } \delta M_2 = \text{some finite value} \tag{i}$$

and

$$\delta M_2 = 0 \quad \text{when } \delta M_1 = \text{some finite value} \tag{ii}$$

Applying (i) to (3), eqns (3) reduce to

$$1 = wp + xr$$
$$0 = yp + zr$$

and applying (ii)

$$0 = wq + xs$$
$$1 = yq + zs$$

which can be solved to give values for $p$, $q$, $r$ and $s$. Thus

$$\begin{bmatrix} p & q \\ q & s \end{bmatrix} = \begin{bmatrix} z & -x \\ -y & w \end{bmatrix} \times \frac{1}{(zw + yx)}$$

The decoupling matrix can thus be formed from the open-loop process gain array:

$$\begin{bmatrix} p & q \\ r & s \end{bmatrix} = \begin{bmatrix} +0{\cdot}062 & +0{\cdot}036 \\ +0{\cdot}061 & -0{\cdot}148 \end{bmatrix} \times \frac{-1}{(0{\cdot}148 \times 0{\cdot}062) + (0{\cdot}061 \times 0{\cdot}036)}$$

$$= \begin{bmatrix} -5{\cdot}452 & -3{\cdot}166 \\ -5{\cdot}364 & +13{\cdot}014 \end{bmatrix}$$

*Operation of Decoupling Matrix*
The following illustrates the way the decoupled control action works. A small change in both $Al_2O_3$ and $Na_2O$ content of the kier liquor has been assumed (one positive and the other negative) and the proportional

control action (assuming unity controller gain) calculated to show that *both* errors are *simultaneously* reduced by approximately the same proportion.

Assume a change in KL composition from design to 168 g/litre of $Al_2O_3$ and 157 g/litre of $Na_2O$. Then NL contains

$$[168 + (0 \cdot 7575 + 88 \cdot 7863)]/1 \cdot 7575$$
$$= 133 \cdot 8453 \text{ g/litre of } Al_2O_3$$

and

$$[157 + (0 \cdot 7575 \times 120 \cdot 1850)]/1 \cdot 7575$$
$$= 141 \cdot 1264 \text{ g/litre of } Na_2O$$

hence the error in $Al_2O_3 = 135 - 133 \cdot 8453 = 1 \cdot 1547$ g/litre and the error in $Na_2O = 140 - 141 \cdot 1264 = -1 \cdot 1264$ g/litre. Then:

$$\begin{bmatrix} \delta C_1 \\ \delta C_2 \end{bmatrix} = \begin{bmatrix} -5 \cdot 452 & -3 \cdot 166 \\ -5 \cdot 364 & +13 \cdot 014 \end{bmatrix} \begin{bmatrix} 1 \cdot 1547 \\ -1 \cdot 1264 \end{bmatrix}$$

$$= \begin{bmatrix} -2 \cdot 729 \\ -20 \cdot 853 \end{bmatrix}$$

expressed as % changes in ratios.

Hence new ratio SW/KL = $(1 - 0 \cdot 027 \ 29) \ 0 \cdot 7575 = 0 \cdot 7373$ and new ratio OL/SW = $(1 - 0 \cdot 208 \ 53) \ 0 \cdot 4485 = 0 \cdot 3550$. Then:

$$Al_2O_3 \text{ in SW} = (0 \cdot 3550 \times 75) + (1 - 0 \cdot 3550)100$$
$$= 91 \cdot 1246 \text{ g/litre}$$

and

$$Na_2O \text{ in SW} = (0 \cdot 3550 \times 145) + (1 - 0 \cdot 3550)100$$
$$= 115 \cdot 9754 \text{ g/litre}$$

Hence:

$$Al_2O_3 \text{ in NL} = [168 + (0 \cdot 7373 \times 91 \cdot 1246)] \div 1 \cdot 7373$$
$$= 135 \cdot 37 \text{ g/litre}$$

and

$$Na_2O_3 \text{ in NL} = [157 + (0 \cdot 7373 \times 115 \cdot 9754)] \div 1 \cdot 7373$$
$$= 139 \cdot 59 \text{ g/litre}$$

and it can be seen that the new error vector is:

$$\begin{bmatrix} 0{\cdot}37 \\ -0{\cdot}41 \end{bmatrix} \text{ as opposed to } \begin{bmatrix} 1{\cdot}1547 \\ -1{\cdot}1264 \end{bmatrix}$$

The percentage reduction in each error is very nearly the same:

$$\begin{bmatrix} 68\% \\ 64\% \end{bmatrix}$$

thus demonstrating that the two control actions are almost totally decoupled.

## 21.3 DYNAMIC PLANT INTERACTIONS

In general the dynamic behaviour of the plant is regulated by the SISO controllers, and only the steady-state or process interactions need to be decoupled as illustrated in the previous sections. In a small number of very important cases the process interactions are more complex and cannot be regarded as steady state only. Two such examples were given in Section 10.13 of this book, both of which involved inverse responses. In the case of the boiler drum level control the conventional '3-term' control system is probably the best solution, but it should not be overlooked that the rate of change of a measured value is itself an independent measured variable. Thus it can be included in the vector **u** when designing a decoupling mechanism, provided that it can be effectively measured. In many cases all that this involves is an additional computation utilising a number of sequential sampled data values of the process parameter in question. It may then, in specific cases, be possible to include in the decoupling algorithm elements of dynamic behaviour. The ultimate in such a concept is embodied in Rosenbrocks theory of multivariable control.

## 21.4 PARAMETERISATION

It is sometimes difficult to measure directly the parameter which describes the product quality. A good instance of this is to be found in 'shape' control of metal rolling mills. Shape is the variation of gauge or thickness across the sheet as it emerges from the mill. Such variation

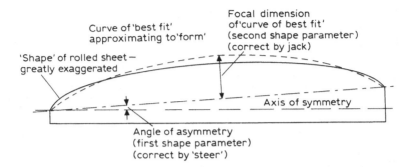

**Fig. 21.5.** Pictorial illustration of shape parameterisation.

stems from the fact that both ends of the rolls are forced down on the moving sheet by hydraulic jacks which do not always apply exactly the same pressure; variation arises also from bending of the rolls themselves (see Fig. 21.5).

Shape has to be defined in terms of more than one parameter, each of which can be directly *and independently* manipulated by the available final control elements. In this case, since there are two and only two final control elements, it is essential that shape be defined in terms of two substantially independent parameters. As can be seen from Fig. 21.5, one such set of two independent parameters can be tilt or 'steer' and the focal length of the 'curve of best fit'. The former is manipulated by changing the ratio of forces applied by the two jacks, whilst the latter is affected by the total force applied, almost independently. Thus a decoupling matrix can be devised and used within the control loop to implement two-variable control, provided also that the measurement of shape can be 'parameterised'.

Shape is measured by a number of gauge detectors/transducers placed at intervals across the emerging sheet. The measured value of each of these transducers taken together provides a profile of the shape of the sheet, as shown in Fig. 21.5. The calculations required to 'extract' the two required parameters from these multiple measured values can be accomplished by a dedicated microprocessor sub-system in the form of discreet sampled data, and subsequently fed to the control system microprocessor for it to calculate new values of the two control outputs.

As one of the control actions required to control shape is the total force exerted by the hydraulic jacks, it follows that shape and mean gauge are interacting variables and cannot therefore be controlled independently on plant comprising a single pair of rolls. More than one set of rolls must be used: this will provide a third parameter, and this in turn will enable shape and gauge to be independently controlled by a system which utilises a decoupling matrix in the control system and parameterises the measured data.

## REFERENCES

Bristol, E. H. (1966). On a new measure interaction for multivariable process control. *I.E.E.E. Transactions on Automatic Control*, January 1966, 43–4.

Rosenbrock, H. H. (1968). Design of multivariable control systems using the inverse Nyquist array. *Proceedings of the I.E.E.E.*, **116**, 1929–36.

Shinsky, F. G. (1967). *Process Control Systems*, McGraw-Hill Co., New York, pp. 131–44.

# CHAPTER 22

# Emergency Shut-Down Systems

## 22.1 INTRODUCTION

The last chapter dealt with the hardware and software structure of the typical instrumentation and control system, and whilst it touched briefly on the subject of reliability, this subject warrants a section of its own. This is particularly true in the case of systems which monitor the safe operation of the plant, as the recent Piper Alpha disaster demonstrated so tragically. In fact, as was pointed out in the last chapter, safety monitoring systems must be entirely separate from process control systems because they have to be designed against a different set of priorities. Whilst some limited 'down time' due to failure of hardware or software is acceptable in most process operations, no failure of any sort is really acceptable when monitoring dangerous plant conditions. However no man-made system can ever be perfect, (and in practice all fall a long way short of this ideal) and it has to be accepted that occasional failures will inevitably occur. This being the case, it is essential that any such failure should be 'fail safe', that the statistical probability of any failure should be reduced to an extremely low figure and that faulty equipment can be immediately identified, replaced and commissioned. In this chapter, therefore, these objectives are analysed and the way that they can be met by the equipment manufacturer and the user is examined.

## 22.2 AVAILABILITY

The term 'reliability' when applied to complex systems is difficult to define. In fact its definition is now an accepted branch of engineering in its own right. However an estimate can be made, in statistical terms, of the 'availability' of a distributed processing real-time system. Mean time between failure (MTBF) of individual components such as resistors, capacitors, transistors or chips, can be evaluated by statistical analysis of tests on multiple-component samples. Military equipment manufacturers in particular have long demanded such information. From this data it is also possible to synthesise a MTBF figure for a total system, and this too has become a contractural requirement by the purchasers of large systems, even in the industrial field. It is however important to remember that such figures are merely predictions of probability, and in no way guarantee that breakdown will not occur at *any* time. 'Availability' can be arrived at from the MTBF and the MTTR (mean time to repair); both parameters that a manufacturer can reasonably be asked to supply.

If the fraction of time that a system will be available is given as $A$ then

$$A = \frac{\text{MTBF}}{\text{MTBF} - \text{MTTR}}$$

Since the fraction of time it will *not* be available is $(1 - A)$, it follows that if redundant equipment is provided with automatic changeover, then in the event of failure of the operating unit

$$(1 - A) = (1 - A_1)(1 - A_2) \ldots (1 - A_n)$$

where $A_1, A_2, \ldots, A_n$ represent the availability of the $n$ systems provided (i.e. $n - 1$ redundant systems). Hence

$$A = 1 - (1 - A_1)(1 - A_2) \ldots (1 - A_n)$$

If, for instance the availability of each of two systems was assessed as 99% (non-availability 1% or 0·01) then the availability of the redundant system is given as 99·99%. This is, of course, only a theoretical probability, and in any case takes no account of certain important facts. For instance, it has been found that under certain conditions the 'shelf life', in store, of electronic components can be *less* than if the component is operating. Hence the concept of 'hot standby'.

The availability of a distributed system means something entirely different from the availability of either a DDC system on the one hand, or an analog system on the other. In the case of the former, failure is almost certain to shut down the entire instrumentation system. Hence many users of such systems considered it necessary to provide full or partial analog back up, making such systems totally uneconomic as a replacement for analog control. In the latter case failure, though relatively frequent, usually involved only a single SISO control loop, and was therefore of no great consequence to the operation of the plant. In the case of the distributed control system, not only does failure of a single non-redundant unit such as a data-gathering/output unit affect a limited number of control loops, but the fact that a standard replacement can be installed and on line very quickly means that a single failure is probably no more serious than failure of a single analog SISO controller used to be, and certainly very much less serious than any single failure of a DDC system.

## 22.3 INTEGRITY

This term is used to express the confidence that the user can have, that the system, whilst not demonstrably in failure, is nevertheless performing *all* its functions correctly. The concept of the 'real-time watchdog', first introduced with DDC, together with extensive on-line diagnostic capabilities, ensures that the integrity of the distributed instrumentation system is at least as good as any DDC system and better than an analog system which does not have the benefit of self diagnostics. It is indeed this last point that constitutes the best reason for development of digital transducers to replace the analog transducers that are still predominantly used. Ideally each processor should have its own independent real-time watchdog which should report to a central on-line diagnostics system.

## 22.4 SECURITY OF SAFETY SYSTEMS

A safety monitoring system must never fail in a manner that causes a dangerous condition; it must be 'fail safe'. In the past, when an individual circuit for each and every condition monitoring device was the normal (and indeed the only possible) practice, it was relatively easy to design each circuit to be fail safe. In the event of an equipment failure, as opposed to an alarm or shut-down state occurring, it was usually simple to determine afterwards what had happened, by testing the offending circuit. In the case of a distributed system in which a multiplexing device 'scans' multiple-fault detection devices, there are no individual circuits! The failure of any sub-system means that large numbers of protection devices, or even all, cease to give the protection intended. The systems may be designed 'fail safe', indeed must be, but even so extensive loss of protection is unacceptable. In addition there is the problem, mentioned in Section 19.10, of detecting 'first-up' alarm or shut-down occurrences, which makes it difficult or even impossible to establish quickly which device caused the alarm or shut-down.

Distributed safety shut-down systems must obviously be designed so that the chance of failure to operate correctly is negligible. As has already been pointed out, any stand-by system must be 'hot' if it is to be relied upon to operate in the event of the primary unit failing. It must also be possible to detect instantly the failure of either the primary or the stand-by unit. The only way to ensure that no fault has occurred in either system is to compare the results of their operations with each other. Whether the operation is transferring data over a communication link, or processing data in some way, this will always detect a fault. However, the question then arises as to which system is faulty. For this reason the technique of comparing the results of operations carried out by two 'hot' systems has to be combined with another technique — 'voting'. Voting entails a logical decision as to which system is most likely to be faulty. However there is no basis for such a decision if only one back-up system is provided, since one system is as likely as the other to be the culprit. An odd number of systems is required and so 'voting' can only be used if at least three systems are provided. The chances that two systems, which are in total agreement in respect of a message consisting of thousands of 'bits', being in error and of a third, which is in disagreement being correct are very small indeed.

Even if every one of the sub-systems is triplicated there will still remain the possibility that the software contains a fault which will give the wrong result under certain (untested?) conditions. Since the same programs will be run in each of the three systems, it is probable that the same faulty result will occur in each. Unfortunately there is no way at present that software can, either by design or by testing, be certified fault free. Thus for maximum security, each of at least three systems will have to be provided with totally independent software which has been designed by separate design teams to meet a common specification. Only then can agreement between two out of three systems be taken to indicate that *they* are operating correctly and that the third system is faulty. Not only is the cost of implementing the above philosophy usually considered excessive, but it has been found impracticable to realise, mainly because the probability of a software fault is increased threefold. Instead, careful design, thorough testing and reasonable 'hot' redundancy, together with fail-safe design must be relied on. This provides a system which is very unlikely to fail, but which, if it does, fails safe and can be repaired very quickly.


## 22.5 REDUNDANCY

Assuming that the MTBF of all components of a sub-system is such that, with a total replacement first-line servicing policy, the system availability (see Section 22.2) is better than 99%, then one stand-by system can provide availability of 99·99%. This is equivalent to saying that the statistical probability is that the equipment will be non-functional for no more than 1 h in every 10 000 (1 hour per year approximately). Component life MTTF (mean time to fail), as measured statistically, is such that this level of availability should be easily exceeded, as in practice it is.

Assuming, therefore, that every sub-system and every communication link is duplicated, that both are operational and that the results of all operations by duplicate systems and communication links are compared to establish 'no-fault' operation, then the system should be capable of providing an acceptable availability. A little thought will reveal that this condition can only be met if *either* of any duplicate sub-systems can communicate with *both* of any other two sub-systems via *both* of the communication links. This provides that each sub-system receives a total of four identical messages under 'healthy' conditions. As shown in Fig. 22.1, it is also necessary that the communication 'handler' in each sub-system be duplicated in order to meet these conditions. Each duplicate handler is associated with only one of the communication links: thus each handler receives two identical messages, one from each of the duplicate transmitting sub-systems. If the messages match, then it is certain that the two transmitting sub-systems are healthy. Provided, then, that each of the two handlers receives two identical messages, comparison can be made next between messages received by each handler. If this shows all four messages to be identical, then not only are the two transmitting sub-systems both healthy, but also both communication links must be healthy *and* free from interference. If the two messages received by the two handlers are not identical, then either one or both of the communication links are subject to interference, or one of

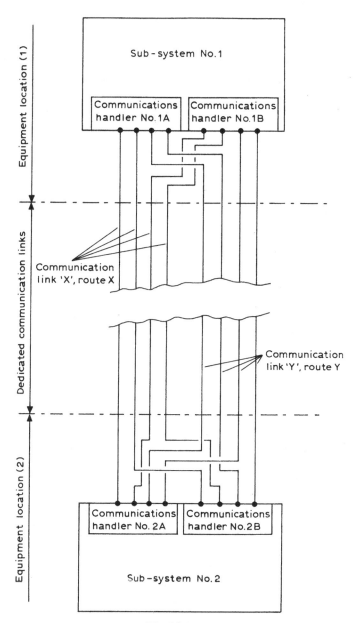

**Fig. 22.1.**

them is faulty. The first response to a mismatch, then, will be for both handlers to request retransmission of the message: consistent failure will be construed as a faulty communication link.

In conjunction with monitoring by appropriate real-time watchdogs, the foregoing philosophy provides the basis of system on-line self diagnostics. A self-diagnostic system is an essential part of any distributed processing safety monitoring system. The structure of a typical sub-system of a safety shut-down system is now apparent, and is illustrated in Fig. 22.1.

Whilst, on the face of it, such a structure meets the requirements of a secure system, the additional time occupied in processing the message comparisons exacerbates the problem of how to detect first-up alarms signals. There is no simple answer to this problem and, where necessary, conventional specialist alarms systems designed to provide this facility will have to be used at the lowest level of the system, instead of simple multiplexed 'scanning'. Such modern alarm systems are invariably constructed from LSI circuit chips and there should be no real difficulty in integrating them into any manufacturer's system.

## 22.6 FAIL-SAFE OPERATION

The design of a system which is fail safe, i.e. will never cause a dangerous condition even when it fails, is not a function of the equipment design, but of the system design. Provided the equipment manufacturer provides the option of contact closure or contact break as the alarm condition to be monitored, it is the definition of the consequences of any monitoring device detecting a dangerous condition that determines if operation is or is not fail safe. The definition of such consequential actions is the responsibility of the user or of the design contractor. However, these definitions are usually taken by the system-design contractor (usually the equipment manufacturer) and converted into software which, when run by the appropriate sub-systems, implements the logic of these consequential actions. If there are faults in this software, so that the required consequential actions do not take place, or faults in the definition of the consequential actions themselves, then the system will not operate in a fail-safe manner (see Section 22.7). Only careful checking of both phases of design, followed by rigorous testing of a completed system, can ensure it provides truly fail-safe operation. Evidence of such rigorous testing will always be demanded by the appropriate statutory organisations.

In addition to establishing that each and every condition monitoring safety device will cause the monitoring system to implement the required fail-safe shut-down action, rigorous testing of the system will sometimes reveal unexpected 'timing' problems which themselves render the system unsafe. The author has personal experience of a very large 'fire and gas' system which was found to take more than 15 s to react to a detection device by generating the required shut-down signals. The explanation lies in the complexities of the system. At the lowest level many binary-state devices have to be scanned to determine all states; the data has to be collected into a data base to facilitate the construction of messages for onward transmission, into the data bases of higher-level sub-systems. These systems use the latest-state data in programs which implement logical relationships to produce a set of output 'states' which are then downloaded to the data bases of output systems. These in turn access their data bases in order to update the state of many operating devices such as shut-down valves, contactors, etc. At each stage there are inevitable delays similar to those described in Sections 19.5 and 19.6. These delays are exacerbated by the complexities of message comparison and checking which are essential to secure operation. It should not be, (though it often is) any surprise that these factors result in overall delays many orders greater than the delays which are to be expected at each stage of the process. Careful design of the software is essential to prevent this sort of problem arising!

## 22.7 SECURITY OF SOFTWARE

As the cost of hardware diminishes in real terms and that of software inexorably increases, the concepts of system design must be continually under review. Software which enables equipment to function is not normally the problem; this can be rigorously tested in the performance of every possible function. The 'application programs' created to implement the user's system logic are another matter: they are unique to a particular system and are usually tested at a stage in the supply contract when there are very real commercial pressures to curtail the time devoted to this work.

# CHAPTER 23

# *The Microcomputer in Process Control*

## 23.1 INTRODUCTION

Chapters 21 and 22 dealt with the real-time control techniques which it is possible to implement using distributed processing equipment based on microprocessors. In this chapter, then, it is appropriate to take a look at control techniques which do not involve real-time data processing. These include 'supervisory control' or on-line process management as well as 'model reference feedforward control' schemes and they constitute a higher level in the hierarchy of process operation.

One of the products of microprocessor technology is of course the microcomputer, which utilises a single microprocessor as its central processor (CPU). The recent introduction of 32-bit microprocessors and true 'multitasking' operating software has brought the wheel round full circle, since these 'desktop' microcomputers can do what, less than two decades ago, required a large mainframe computer. Such computers, or if necessary larger computers, can readily be interfaced with the common data highway of a distributed control system. In fact there is no reason why a number of such computers cannot 'sit' on the system highway giving a number of line managers access to all the available measured value data, and also enabling them to download new operating set points to the system controllers in order to optimise process operation in some sense. The real-time controls automatically regulate the operation of the process, maintaining process set points such as temperature, pressure, level or composition at values determined by management as those which are currently optimal; management can now have the facility to change these set points to other values as a result of computation carried out (in non-real time) by desktop microcomputers which have immediate and automatic access to the latest measured value data base of the distributed control system.

There are a number of ways that computers operating in non-real time can contribute to optimal operation of the process and plant. It is not within the scope of this book to develop these concepts in detail, but it is appropriate to introduce them to the reader in such a way that he will understand their place within the hierarchy of control and operation management which has been made possible by the development of the microprocessor and associated technology. If the reader wishes to study any of these techniques in detail, there are many good books on each of them.

## 23.2 OPTIMISATION

Optimising procedures comprise the management techniques which are used to calculate set-point data which will provide optimal product quality or throughput or meet any other objective set by management. A discussion of those optimising techniques which may be employed is beyond the scope of this book. Such techniques included 'linear programming' for constraint control, 'hill climbing' and a variety of statistical methods.

## 23.3 PROCESS AND EQUIPMENT MODELS

Figure 23.1 shows one possible system structure for optimisation. Each of the five blocks in this diagram can be implemented by a single, or more likely by different, processors in non-real time.

Many books have been written on the subject of mathematical models, which essentially consist of sets of equations which describe the behaviour of the systems or plant in appropriate terms. Thus a process model will comprise a set of equations which describe the way the chemical or other nature of the process behaves. Similarly the operation of the plant can be modelled by sets of equations.

Process and equipment models can be constructed and programmed to run in non-real time, using the latest measured values from the distributed control system. For continuous processing plants which operate essentially in steady state, operation of such models can provide predictions as to the yield or quality of the product at some future time,

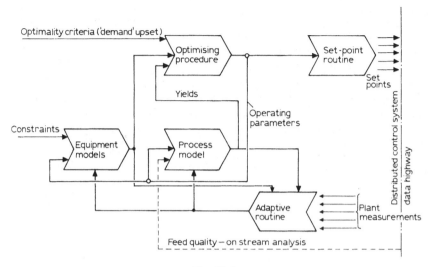

**Fig. 23.1.**

enabling the operator to make corrective changes now to the set points of some controllers for instance, in order to achieve better results in actuality. However, the operator is rarely in a position to calculate what changes of set point are necessary to achieve such future improvement in operation. An optimising procedure is usually required for this purpose and this too can be programmed and run in non-real time, with predicted value data, on an entirely different computer. This computer is likely to be found in the process manager's office, since operation strategy is his function. In addition to the predictions of process and plant behaviour over some finite future operating period, the process manager will need to input data concerning any known changes in feedstock quality or availability. He may also need to be able to change the criteria of optimality, reflected in the program his computer is executing, from time to time.

Neither plant nor process models can normally be 'stationary'; that is to say that heat transfer rates, for instance, will change as surfaces become dirty or are cleaned. This sort of change affects the constants in the equations which make up the model. By comparing certain actual measurements with predictions over an appropriate period of time it is often possible to build adaptive routines into modelling programs: this will enable these constants to be gradually changed in line with actual changes in plant or process behaviour. For instance, in a catalytic reaction process, the equations representing the chemical reactions and kinetics which comprise the process model will incorporate a factor representing the amount of catalyst which is active at any time. This may change with time, and comparison of the calculated outputs of the model with the actual measurements from the plant will allow this factor to be adjusted so that the model yields correct results. The same model (running in non-real time faster than the process) is also used to give predictions of the process outputs, this information being in turn used to generate set point changes which will bring the actual outputs closer to the optimum. The model is being used in a feedforward control sense and is at the same time being adapted to correct for non-stationary behaviour. Similarly a model of the reactor heat-transfer kinetics can be adapted to correct for changing heat-transfer coefficients as the heat-transfer surface becomes dirty or fouled.

All these functions can be connected up in closed loop, as shown in Fig. 23.1. However it must not be thought that the design of such 'model-based' control systems is simple: the model reference feedforward control and the adaptive routine are both multivariable closed loop feedback paths into systems which are invariably non-linear. Mathematicians have recently discovered that even simple non-linear feedback systems often behave in a highly unpredictable fashion. Such systems and the mathematical models on which they are based have to be subjected to rigorous testing and development in the field before they can be commissioned fully. Nevertheless, there is undoubtedly a place for such control schemes, and the equipment systems necessary to implement them are available now.

## 23.4 BATCH PROCESS CONTROL

Process/plant models and optimising procedures can be used in a rather different way for batch processing control where, as not infrequently is the case, a parameter such as temperature of reaction is to be varied according to a strict regime throughout the batch cycle (e.g. manufacture of polystyrene). The product quality may be very seriously affected if the regime is not faithfully followed with each batch; large quantities of scrap material may be generated if reaction temperature departs from the optimal regime. In such situations models can be used to predict the change in the parameter of interest which will take place

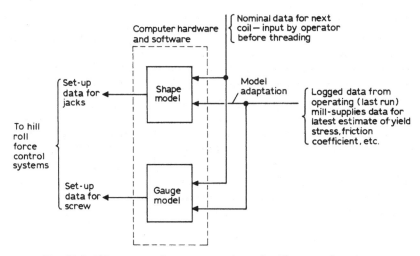

**Fig. 23.2.** Diagrammatic representation of mill set-up functions.

over a relatively small period of time starting with the current set of measurement data, and assuming that rates of heat removal from or addition to the process continue 'as set'. The predictions obtained can then be used to compute new set points for controllers of these variables in order to correct any 'under or over shoot'. This strategy is really a form of feedforward control and will be combined with feedback regulation based on the error that does develop due to any inaccuracies in the models and the control actions.

Adaptive techniques could be used for the batch process also, restricting changes in the models to the beginning of each batch, basing the adaptation on comparisons made during the preceding batch, as illustrated for the case of rolling mill control in Fig. 23.2.

## 23.5  BUILDING MANAGEMENT SYSTEMS

Heating and air-conditioning of buildings (which include factory buildings) have long been considered a separate branch of instrumentation and control. This has more to do with commercial considerations, because the equipment is in a lower cost bracket and must therefore be designed and constructed to different specifications. Nevertheless the complexity of such systems is just as great as that of industrial control systems, even if the variety of problem is considerably less. It is not surprising to find, therefore, that distributed processing has made its impact in this field, just as it has in the industrial control-system field.

The initial stimulus for applying distributed processing to building plant control was energy conservation. The flexibility of such systems made it easy to program 'optimum start and stop' into the boiler-running schedules, for instance. It was then a small step to making these optimising programs adaptive. Software was introduced which, using historical data relating to daily weather conditions, room occupation, etc., enabled the system to vary the optimum start and stop times and thus make further energy savings.

# CHAPTER 24

# *Expert Systems*

## 24.1 INTRODUCTION

In recent times the price of microprocessor hardware has been halved about every 4 years, whilst the cost of writing 'software' to run on the hardware continually increases. As a result, the cost of the software for any particular application is likely today to be greatly in excess of the cost of the hardware or equipment. Because of these fundamental changes, the approach to any engineering system design which is based on the use of microprocessor equipment, must be continually reassessed. On the one hand much more complex hardware can be afforded, whilst on the other hand there is a very strong incentive to find a way to use standard, rather than specially written software for each application. Research in this area has primarily focused on the development and application of new programming tools with which to transfer more of the work to the hardware. So-called 'fifth' generation programming languages have been developed, and it has been claimed that they will result in little short of a second industrial revolution. The Japanese in particular have taken this very seriously; so much so that our own government instituted the Alvey project to evaluate the impact of these developments on industry in the medium term.

Conventional computer languages merely extend the basic concept of giving commands to the processor; high-level commands in languages such as 'C' are themselves programs, comprising many assembly level commands. These languages differ mainly in the way that the high-level commands are put together and compiled. Using this type of language the programmer must define *exactly* the procedures which the processor will carry out in order to achieve the objective/s of any given task. Such programming languages are called 'procedural' languages. Fifth-generation or 'declarative' languages require only the rules of the application to be defined by the programmer: the compiler defines the procedures for the processor to execute: this it does by relating logically the rules and data available.

## 24.2 PROCESS CONTROL

There is considerable potential for expert systems in process control where the equipment is required to make decisions as opposed to being programmed to carry out a defined and unvarying set of instructions. Batch sequence control systems constitute one obvious area in which decision making by the control system could offer very considerable advantages. For instance, the operating strategy for a water supply company which obtains its water mainly from bore holes is a complex of rules concerning pumping rates, water level conditions, etc. An expert system, developed in the first instance for the human operator to consult before taking any action, could eventually be incorporated into a closed-loop system which would relieve the human operator from all but the most difficult decisions. Other areas of control in other industries which involve scheduling of operations, such as despatching in the pipeline industry and combined process and heating schemes are obvious examples.

## 24.3 BUILDING ENERGY MANAGEMENT SYSTEMS (BEMS)

In recent years the distributed instrumentation system has found an application in the building industry in the form of 'building energy management' or more accurately 'building management' systems. Control of heating and ventilating (HVAC) systems for both humidity and temperature, using psychrometric relationships, is a complex matter: expert system software can be used to provide optimising techniques in this area, for instance. Recently, distributed data processing has been applied to systems which monitor the temperature, occupancy and security of individual rooms. It is easy to see the potential for rule-based expert system software to relate the data obtained from such monitoring systems to the management of energy saving, intruder security and fire/smoke protection in designing future building management systems.

## 24.4 SAFETY SHUT-DOWN SYSTEMS

Emergency shut-down systems for North Sea oil/gas rigs have evolved from earlier versions which were implemented in hardwired relay logic. Modern systems, using distributed microprocessor technology, are far more flexible and reliable, but their performance is still far from satisfactory. The main problem lies in the unsatisfactory design interfaces which are implicit in current practice and which stem from the nature of current technology.

The 'application software' of a safety shut-down system comprises the logic relating the states of large numbers of detection devices to the required states of a large number of shut-down devices such as motorised valves, electrical contactors and motor starters, etc. Conventionally, the user or design contractor defines the rules or logic of the shut-down system as an engineering drawing, in the form of a matrix of columns and rows which represent the inputs from detection devices and the outputs to shut-down devices, respectively. Programming is sub-contracted to the equipment manufacturer, who uses the main design contractor's 'matrix diagrams' as his specification. Considerable difficulties are invariably caused by this design interface, if only because the matrix diagrams defining the system logic are subject to continual revision by the main contractor during the design phase. This gives rise to many charges for corresponding revision of the sub-contractor's design. Of even more importance, particularly since the Piper Alpha disaster, is that the complexity of the software-based logic of the completed system is sometimes such that the system response time is totally unacceptable. Important requirements, such as 'first-up alarm detection', complicate the software design and exacerbate this problem. The writer has personal experience of one such system which, when tested, took 20 s to respond to input status changes in certain circumstances.

It is not really possible to devise a testing procedure which can be guaranteed to discover *all* the 'bugs', and it is therefore always possible that some fault will occur, due to the software rather than the hardware, during normal operation. It has been proposed that three identical systems should be provided, together with a 'voting' system: if a fault should arise in the hardware or software of any one system the 'majority' vote of the other two would be acted upon. Three 'sets' of application software would be produced, completely independently, to run on the three separate hardware systems. The same software fault is unlikely to arise in each system, and hence this triplication of systems should greatly reduce the possibility of an undetected software fault causing an unnecessary shut-down, or worse a failure to make a *necessary* shut-down. However, adoption of this idea would not only triple the already very considerable cost of producing the application software, (to say nothing of the cost of triplicated hardware) but would also greatly increase the cost and time taken to implement each and every design change, of which there are always many during the design of such systems. Each system would statistically have the same chance of containing undiscovered 'bugs' which are inevitable in the 'procedural' type of software current today. As a direct result there would be a much greater likelihood of a fault arising in any of the systems due to software malfunction. It has indeed been demonstrated, in the case of the triplicated systems used in the American space shuttle, that this idea is totally impracticable. It produces a system which, though it may 'fail safe', is unreliable, generating too many unnecessary shut-downs.

The main design contractor must define the logic, whichever type of language is finally used to incorporate it into software. Using a declarative language, however, the process of writing the programs to implement these rules could be made much simpler. However, because the process of translating the rules of the system into 'application software' is simpler, the compilation of the program into assembly-level instructions, which the processor can execute, is correspondingly less simple. One is relying on the expertise of the 'software house' that constructed the language environment to ensure that there are no 'bugs'. In other words, much of the onus for designing good 'bug-free' software is shifted from those who program each individual application, to those who construct and market the language compiler.

An emergency shut-down or fire and gas system for an offshore rig will in effect ask a cycle of 'questions', which the software then answers by reference to a data base comprising the current status data for a very large number of sensors and instrument transducers. This data is updated by a data acquisition system which operates at a speed commensurate with the response required for safe operation. These 'questions' concern the appropriate state of control elements such as electrical contactors, shut-off valves, etc.; the 'answers' are then actioned. The 'questions', programmed in a procedural language, constitute the 'application software' of the system.

Using a declarative programming language, the resulting system would be radically different. There would be three data bases: the first, comprising the rules of the application, would be constant; the second, as in existing systems, comprising current status data from the sensors and instrument transducers. At each successive iteration of the cycle of questions, the system would deduce the answers by reference to the latest state of the sensor data, and the rule or knowledge base. The resulting current status of the answer to each such question would then be used to update a third data base, which in turn would be scanned by an executive system, which would shut down electrical and other equipment as and when appropriate. Thus both the hardware and the standard software of the system could be modularised, and the extent of application programming required to implement any particular system could be considerably less than in current systems programmed piecemeal in procedural languages.

Such an approach to the design, by contractors, of an emergency shut-down system could have several advantages. The possibility of undiscovered 'bugs' in the software should be greatly reduced, as should the time taken to complete and test the design. It should also be much easier to maintain, extend or modify the software during the design phase and throughout the life of the system, once commissioned. Perhaps most important of all, the operation of the completed system is likely to be much more predictable, since it would depend more upon standard than on specially written software. It is possible that the application programming, as is currently the case for distributed instrumentation systems, could be simplified to the extent that the main design contractor could undertake this work. If so, then the design interface, cause of so much trouble, would no longer exist. ·

It is not obvious how declarative languages could or should be used to obtain these advantages; indeed it has yet to be demonstrated that these advantages *can* be realised. There is a fundamental option either to produce an operating system, programmed in one of these languages, or to use an expert system 'shell' as the basis for 'application building'. Available expert system 'shells' are however produced for 'off-line' types of application such as medical diagnostics or engineering design. How such systems would perform in an on-line automatic role would have to be assessed.

[In this context it is interesting to note that one such shell, which is reputed to have captured more than 50% of the total UK market, was originally written in such a language, but has been rewritten in its latest version in a conventional procedural language.]

# Index